AF553392

UNDERSTANDING CRYOBIOLOGY

By

Dr. G.P. Garg

&

Dr. M.Prakash

DISCOVERY PUBLISHING HOUSE PVT. LTD.

NEW DELHI-110 002

First Published-2010

ISBN 978-81-8356-515-8

Published by:

DISCOVERY PUBLISHING HOUSE PVT. LTD.
4831/24, Ansari Road, Prahlad Street
Darya Ganj, New Delhi-110002 (India)
Phone: 23279245 • Fax: 91-11-23253475
E-mail: parul.wasan@gmail.com
info@discoverypublishinggroup.com
Website: www.discoverypublishinggroup.com

Printed at:

Sachin Printers
Delhi

Preface

The present title "Understanding Cryobiology" has been written for those students interested in careers in diverse fields of biological sciences. It provides a structured approach to learning by covering all the important topics in a uniform, systematic format. The book has been comprehensively designed incorporating recent advances in this fast moving field. It also provides accessible information on cryobiology in compact form for undergraduate students in biology and related life sciences. It is intelligible to the educated layman, though it deals with some complex ideas. It is an adequate text for all the requirements of students in this area. In addition, busy lecturers who require a quick reference compendium will find it useful, particularly for tutional planning. Simple, yet hopefully clear figures and tables are provided throughout the book.

The over-riding goal of this book, and indeed of the whole *Understanding series*, is to present the essential information concering cryobiology in a compact, readily accessible form which leads itself to student learning and revision. The convergence of various approaches has generated a rich panorama of detail, the significance of which we are still attempting to unraval. The present text has been written as an introduction to this rapidly growing field.

To make the work more comprehensive and informative, the author has consulted many authoritative books, research journals, abstracts, monographs etc., so there can be no claim to originality except in the manner of treatment.

The author expresses his thanks to his friends and colleagues whose continue inspirations have initiated him to bring out this book.

The author expresses his gratitude to Mr. Wasan and staff of M/s Discovery Publishing House Pvt. Ltd. for their whole hearted co-operation in the publication of this book.

In the mean time, the author will remain sincerely responsible for any shortcomings of the book and be grateful to the readers for their suggestions and constructive criticism for the continuous betterment of the book. He takes this opportunity to appeal to the readers to send their suggestions straightaway to his Publisher.

Author

CONTENTS

1

INTRODUCTION

Cryobilology is the study of life at low temperatures. The knowledge generated by numerous researches is scattered among the journal publications and conference proceedings. The aim of this book is to gather it to get, at least in grosso modo, a general view of the whole field. The book is primary intended to serve as a guide (or, perhaps, as a warning?) to the newcomers. Still, it is hoped that it will be also useful for those who actively work in this very wide field.

The problems encountered in cryobiology could be crudely assigned to the different levels. Certainly these levels are interacting and thus only multiscale simulation with the two-way information flow could adequately capture all the processes. At present some of the problems seem to be the most critical.

There is an enormous pool of data on the animal adaptation to cold and its molecular basis, but this information is spread across a range of taxonomically unrelated species; thus, one of the current cryobiology problems is to organize this information - extract common mechanisms within and between species and different taxonomic groups and finally to get "a bigger picture".

A similar related problem is the structural diversity of antifreeze proteins. The AFP types are radically different in their primary sequences and 3D structures and yet they all bind to ice crystals and depress the freezing point.

Microscopic or multiscale simulation of the ice propagation at the cellular level is needed to resolve the remaining uncertainties on the mechanisms of freezing in vitro and in vivo. On the macrolevel, the most demanding is the problem of the adequate goal function that could be used in clinical practice for the computerized cryosurgery planning and optimization. Most probably, the construction of such function will require novel models of the tissue thermal properties that could account for the effects of the freezing injuries occurring during the operation.

Table 1.1. Basic physical problems of cryobiology

Biological object	
Subcellular objects	Interaction of cryoprotectant molecule with membrane
	Evolution of membrane and cytoskeleton
	Destruction of membrane
	Ice propagation through the pores
Isolated c ll	Heat mass transfer to (from) environment
	Interaction with ice front
Uniform tissue	"Effective" thermophysical properties
	Heat transfer
	Ice front propagation
	Ice recrystallization
	Microcirculation effects
System "tumor +	Global heat transfer in the system
Surrounding tissues +	Ice fronts interactions
Cryoprobes +	Mechanical stresses
Heater probes"	Optimization of cryoaction protocol

History of Cryomedicine

Origins of cryobiology could be traced down to ancient Egyptians; probably the first scientific account of this science is the monograph published by Sir Robert Boyle "New Experiments and Observations Touching Cold, Or, An Experimental History of Cold, begun. To which are added An Examen of Antiperistasis, And an Examen of Mr. Hobs's Doctrine about Cold. Whereunto is annexed An Account of Freezing, brought in to the Royal Society, by the learned Dr. C. Merret, a Fellow of it. Together with an Appendix, containing some promiscuous Experiments and Observations relating to the precedent History of Cold".

The effect of water supercooling was discovered by Daniel Gabriel Fahrenheit in the beginning of the eighteenth century. Experiments on the freezing of cells started in the end of eighteenth century. In 1776,

Spallanzani noted that spermatozoa, cooled in snow, became inactive but were revived on warming.

In the middle of the nineteenth century, Julius von Sachs developed a low-temperature stage for his microscope that allowed him to detect the dehydration of cells in the presence of the extracellular ice. Hans Molish was the first to study in the end of the nineteenth century the formation of the intracellular ice and to measure the cell volume change.

Experiments in the middle of the nineteenth century by Michael Faraday who achieved a temperature of 163 K by mixing solid carbon dioxide and alcohol under vacuum, the development in 1877 by Cailletet of France and Pictet of Switzerland of adiabatic expansion systems for cooling gases and in 1892 by Dewar of Great Britain the first vacuum flask for storage and handling of liquefied gases as well as invention in 1895 by Linde of Germany and Hampson of England continuously operating air liquefiers based on the Joule–Thompson effect provide the technical support for the cryomedicine development.

In 1833, Openchowski reported using a low-temperature system for freezing portions of the cerebral cortex of dogs, however, not for therapy but as a tool for inducing lesions in the brain for research.

Twentieth century witnessed a rapid development of cryobiology related to the progress of the cryogenic equipment (closed systems based on liquid nitrogen evaporation or on the Joule–Thomson effect – cooling by rapid gas expansion from a high to a low pressure zone through a small orifice), developments of monitoring techniques, extension of the list of diseases that have been successfully treated by cryomedicine, and consolidation of research by foundation (simultaneously in 1964) of two major scientific societies in this field – The Society for Cryobiology and The Society for Low Temperature Biology.

Vitrification as the preservation method was suggested by the Jesuit priest B.J. Luyet in 1937 and demonstrated by him (in collaboration with R. Hodapp) a year later for the frog spermatozoa vitrified in liquid air after dehydration in the solution of sucrose; the fundamental monograph "Life and Death at Low Temperatures" by B.J Luyet and P.M. Gehenio was published in 1940. The authors described extensive experiments on the freezing of a wide range of organisms from bacteria and nematodes to tissues isolated from higher organisms, including frog muscle and mammalian erythrocytes. Luyet and Genehio concluded that frequently cellular damage and death were caused by the ice formation.

In 1949, Polge et al. reported the first use of glycerol as a cryoprotectant. This discovery has started a fast development of cryopreservation techniques and cryoprotective solutions. A cryopreservation was reported for the fowl spermatozoa suspended in a glycerol–albumen solution surviving freezing to –79 °C and later for the bull spermatozoa. Establishment of human sperm banks suggested in 1866 by Mantegazza was realized a century later almost simultaneously in France and USA.

Human embryos are now routinely cryopreserved as an adjunct to other techniques of *assisted reproduction technology* (ART), but still this approach remains the secondary one: babes born from cryopreserved embryos represent less than 8% of the total of ART babies born.

CRYOSURGERY

James Arnott of Aberdeen was probably the first to apply the extreme cold locally for anesthesia as well as for the pallation of tumors in the breast and the uterine cavity and the destruction of the tissue using a mixture of salt and crushed ice. He used a water cushion cooled by the flow of a solution from a reservoir of brine. The cushion was applied to accessible tumors, which were frozen to temperatures of about –24 °C with no hemorrhage. His work was published by J. Bennet of Edinburg in his book "Cancerous and Canceroid Growth" (1846).

Campell White of New York and William Pusey of Chicago were the first to use for treatment cotton swabs dipped in liquid air and solid carbon dioxide (carbon dioxide snow or carbonic acid snow), respectively. C. White used liquid air for the treatment of a large range of diseases, including lupus erythematosus, herpes zoster, chancroid, warts, varicose leg ulcers, carbuncles, and epitheliomas. W. Pusey recognized the low scarring potential of cryosurgery although he attributed this to regeneration of residual epidermal cells rather than to collagen's resistance to cold.

After liquid oxygen became commercially available, it began to be used in the treatment of skin diseases in 1929. However, since liquid oxygen is a fire hazard, it has never become a popular cryogen for cryosurgery. In the early 1940s, P.L. Kapitsa in the SU and Collins in the USA began developing commercial techniques for large-scale liquefaction of hydrogen and helium, with liquid nitrogen as an abundant and low-cost by-product. Soon after liquid nitrogen became readily commercially available, Allington introduced it in clinical practice in

1950. The liquid nitrogen was applied with a cotton swab, and soon became common in the treatment of verrucae, keratoses, and various non-neoplastic lesions.

During the early part of the twentieth century cryosurgery was primarily used to treat dermatologic and gynecological diseases. Treatment of deep tissues was pioneered by a neurosurgeon Temple Fay of Philadelphia using a self-made closed system in 1938 for the treatment of advanced carcinoma, glioblastoma, and Hodgkin's disease. M.J. Gounder and collegues were the first to use cryosurgery for urological disorders in 1960s.

A number of cryogens were used for medical applications, including dichlorodifluoromethane, with the boiling point -29.8 °C, solid carbon dioxide (-79 °C), liquid nitrous oxide (-88.5 °C), liquid air, and liquid nitrogen (-195.8 °C).

Cryosurgery Equipment

Sometimes no cryosurgery equipment at all is necessary: for the treatment of the open tissues such as the malignant bone tumor the so-called direct pour (open system–pouring liquid nitrogen through the stainless-steel funnel into the tumor cavity) method could be used. However, even in this case exploitation of the technological advances (pressurized spraying of liquid nitrogen, a semi-open system) is preferable, since it allows to greatly reduce various complications (skin necrosis, delayed healing, nerve injury, etc.) due to the rapid evaporation of liquid nitrogen from the cavity.

Probably the first cryomedicine system was designed by J. Arnott in the middle of the nineteenth century. It consisted of a waterproof cushion applied to the skin, two long flexible tubes to convey water to and from the affected part of the patient, and a reservoir for the the ice/water mixture with a pump. Using salt/ice mixture, Arnott achieved temperature of -24 °C. For this device Arnott was awarded a medal at the Great Exhibition of London in 1851.

Various techniques were used to administer cold, including precooled metal blocks, precooled needles, dry ice applications, thermoelectric methods, and cryogenic heat pipes. Solid copper cylindrical disks were cooled by immersion in liquid nitrogen before application to the skin. These cylinders by far exceed the cotton applicators in thermal capacity and heat exchange characteristics. Additional effect follows from the pressure exerted on the lesion that reduces the blood flow and related heat source. This approach allowed to double the depth of the tissue destruction to 4–5 mm.

A spray for cryosurgery was invented by a dermatologist H. Whitehouse of New York in 1907. He studied the effects of liquid air on normal skin and treated epitheliomata, lupus erythematosus, and vascular naevi. J.T. Bowen and H.P. Towle reported the successful use of liquid air for vascular lesions in 1907. Liquid nitrogen spray developed by Torre in 1965 was later modernized by S. Zacarian, who suggested the interchangeable tips allowing variation of the spray diameter and designed a hand-held device.

Close systems based on liquid nitrogen were introduced by I.S. Cooper and A. Lee in 1961. Low temperature was achieved by evaporation of liquid nitrogen in a hollow insulated metal probes close to the tip attached to a circulatory pump. Liquid nitrogen transport to the tip is provided by the use of the so-called *Leidenfrost flow* (droplet film-boiling flow). The heat transfer from the liquid nitrogen to the tube wall in Leidenfrost flow is very low because liquid droplets are separated and thermally insulated from the wall by gas, thus little coolant is boiled off during transport to the tip.

A totally closed system allows the surgeon apply the cold to any part of the body accessible to probe. Later Cooper modified his original cryoprobe by adding a heating element that facilitates probe removing from the tissue after freezing, relieving the surgeon from the need to use special tricks such as flushing with saline to assist the detachment of cryoprobe tip from the frozen tissue.

Liquid nitrogen-based cryoprobes provide the lowest temperature of the probe tip (–196°C that could be lower down to –209°C if supercooling of liquid nitrogen is used). They have, however, two drawbacks. First, the diameter of the cryoprobe, due to a rather complex construction providing the efficient two-way flow of nitrogen, could not be made smaller than approximately 3 mm that limit the number of probes that can be used simultaneously (usually 5–6). A small number of cryoprobes forces operator to reposition cryoprobes during the operation to create a composite volume of overlapping spheres (ellipsoids) of frozen tissue. Second, liquid nitrogen systems react slowly to the changes in user input. An experimentally determined time delay is about 1–2 min. In addition, transfer tubes and their associated valves should be precooled before operation; the cryoprobes for the short operations that have a small storage Dewar incorporated onto the probe itself are free of this minor disadvantage.

Modern cryoprobes using the Joule–Thomson effect – a constant enthalpy expansion of gas rapidly cooling it down to the boiling point

– have a significantly smaller diameter. The modern systems based on the ultra-thin gauge needles (down to 1.47 mm diameter) allow direct transperinal insertion and, if necessary, relatively easy reposition of probes during the operation, reducing the disturbances of the tissues. The large number of cryoprobes used simultaneously (up to 17) provides the more flexible control of the shape and extension of the freezing zone, allowing treatment of irregularly shaped tumors. Additional flexibility of the treatment results from the ability of independent modulation in time of each of the multiple simultaneously operated cryoprobes.

Since some gases such as helium under the Joule–Thompson effect warm up rather than cool when expanded, it is possible to create systems for both cooling and heating with the rapid conversion between regimes. The gas cryoprobes using argon and helium can be modulated between −186 °C and +40 °C in about 30s. Different mixtures of hydrocarbons and synthetic refrigerants with argon and nitrogen are also considered as potential working fluids.

Liu et al. developed on the novel minimally invasive probe system capable of performing both cryosurgery and hypertermia. In contrast to the systems using the Joule–Thompson effect, in the new system one can alternatively supply the cryoprobe tip with either liquid nitrogen or hot water vapor. The drawback of the system described by the authors is a rather high diameter of the probe (5 mm) needed to provide transport of a large amount of the working fluid.

Rabin and Shitzer have modified the conventional cryoprobe by adding an electric heater to develop the computer-controlled system that provide such temperature of the probe that a specified constant cooling rate at the crystallization front is maintained.

Another advance of modern cryosurgery systems is the simultaneous use of a warming catheter ("cryoheater") to protect important organs in the close vicinity to the frozen malignant tissues such as the urethral mucosa.

A special cryoprobe for the study of early stages of osteonecrosis by introducing cryo insult in animal models that mimics naturally occurring osteonecrosis lesions was design by Reed et al., which in contrast to practically all commercial cryoprobes based on the "passive" vacuum tube insulation incorporates "active" resistive electric heating that emulate insulation, giving two independent operator-controlled parameters: the mass flow rate of liquid nitrogen and the heating coil current.

In the early years of cryosurgery monitoring was essentially reduced to the temperature measurement, and a platinum resistance thermometer designed by Pegg and Hayes for the temperature range from 37 °C down to -200 °C with an accuracy of 0.1 °C was an essential achievement. Now cryosurgeon could use a number of advanced techniques such as US and NMR imaging.

Evidently, however, the progress in the development of the cryosurgery systems and the operation cooling/thawing protocols could neither diminish the crucial role of the surgeon nor liberate she/he from the responsibility for the treatment. New cryosurgical tools along with modern intraoperative monitoring techniques and mathematical modeling (preferably using the patient-specific anatomical data) just help the surgeon to make right decisions.

Low Temperature Preservation

The first studies in low temperature preservation were motivated by the development of *artificial insemination* (AI) procedures. Beginning of AI is attributed to Arabs stealing stallion semen from a rival tribe. Horses also were the subject of the modern development of AI started by Ivanov in Russia in the beginning of twentieth century. He studied AI in other domestic animals (dogs, rabbits, poultry) as well. Ishikawa, working with Ivanov, began a similar program in horses after returning to Japan in 1912. At the same time, work on AI in horses was done in Denmark, at The Royal Veterinary College in Copenhagen, by Eduard Sorensen, who also was familiar with Ivanov's work. Sorensen is also known for his invention of straw for packaging semen.

In 1938, Jahnel reported the survival of human spermatozoa stored at the temperature of solid carbon dioxide. The systematic use of the frozen-thawed semen techniques was pioneered by Christopher Polge in the late 1940s, who worked with chicken and bull semen. The early preservation procedures were developed in ad hoc manner and consist, for example, in the preliminary keeping the semen in glycerol for a very long time (glycerol equilibration time) or dilution in the sucrose–egg yolk mixture and vapor freezing as pellets.

Sperm motility was, and essentially remains, the main criterion to assess the success of the freezing procedure. Polge was the first to show that too large concentration of glycerol in the liquid semen decreases the fertility by the damage of the sperm cell membrane. The first mammal - a calf - was produced with the cryopreserved spermatozoon in 1951; the first pregnancy obtained with the frozen stallion semen was reported by Barker and Gandlier in 1957.

First experiments on the cryopreservation of human spermatozoa were reported in 1940 by Shelters, who used direct plunging of thin-walled capillars with undiluted semen into alcohol cooled down to -79 °C with solid carbon dioxide, into liquid nitrogen at -196 °C or into liquid helium at -269 °C. First pregnancy from a cryopreserved human embryo was reported in 1983.

A significant progress was achieved by Lovelock and Bishop by discovering the cryoprotective properties of dimethylsulfoxide in 1959.

Probably the most important biological object subjected to low temperature preservation is the human erythrocyte - red blood cell. There are different reasons for the RBC transfusions - large RBC loss (traumatic or surgical hemorrhage), decreased bone marrow production, defective hemoglobin, and reduced RBC survival (hemolytic anemias). Low temperature preservation of RBC ex vivo for clinical use allows to separate the donor and the patient in space and time. Two obstacles to successful transfusion are blood clotting and in vitro loss of RBC viability and function. To overcome them, citrate as an anticoagulant was introduced by Hustin in 1914 and glucose as a preservative by Rous and Turner in 1916; cryopreservation of RBC was pioneered by Smith and Lovelock in the early 1950s . Later both vitrification and lyophilization also have been considered as the preservation approaches for RBC; the latter having evident advantages due to the stability of the dried cells at room temperature for the long time periods.

Advances in modern cryopreservation procedures could be divided into two directions:

1. Optimization of the cooling/thawing protocols, including vitrification-based approaches and development of special procedures
2. Improvement of the cryoprotective solutions

There is a number of special carriers or vessels developed for the CP, such as the open pulled straws (OPS), the flexipet-denuding pipette (FDP), microdrops, hemistraw systems, cryoloop and others.

CP procedures

In total, seven vitrification-based procedures used for the conservation of genetic resources could be distinguished:

1. Encapsulation–dehydration - cells are encapsulated in alginate beads kept for several days in the concentrated sucrose solution; encapsulation–dehydration could be used with two-step cooling method

2. Vitrification – dehydration of sample with highly concentrated vitrification solution and subsequent rapid freezing
3. Encapsulation–vitrification – alginate beads are preliminary dehydrated at 0 °C before plunging in liquid nitrogen
4. Pregrowth – cultivation of samples in the presence of cryoprotectant, then direct immersion in liquid nitrogen
5. Pregrowth–desiccation – dehydration of the samples pregrown in the presence of CPA under the airflow cabinet or with silica gel, then rapid freezing
6. Desiccation – probably the simplest procedure – direct immersion of the dehydrated samples into liquid nitrogen; sometimes the ultra-rapid drying in a stream of compressed dry air is performed
7. Droplet freezing – small drops of the solution with samples placed on aluminum foil and frozen by direct immersion in liquid nitrogen.

Classical cryopreservation involves slow cooling down to a given prefreezing temperature, followed by rapid immersion in liquid nitrogen. In vitrification-based procedures, cell dehydration is achieved before freezing by exposure of samples to concentrated CPA solutions and/or by air desiccation and the subsequent rapid cooling. The critical stage in the classical approach is the freezing step and in vitrification it is the dehydration step. Vitrification has economic advantages, since it is relatively simple, it does not require the expensive programable freezing equipment and is very fast (this procedure requires several seconds).

A multistep freezing process consists, as a rule, of initial slow freezing followed by fast freezing to reach the final low temperature. Such procedure provides preliminary dehydration of tissues or organs that is beneficial in two aspects:

1. It minimize the amount of ice that can grow within organ capillaries, reducing the risk of injury due to the ice expansion within these vesselssuch damage is a known problem in cryosurgery
2. It increases the concentration of the CPA within cells

Multistep freezing is also preferable with respect to the development of the thermal stresses that may cause fracture in the tissues.

A specific method of the oocyte and embryo cryoconservation was reported by Nagashima et al., which consists of the preliminary stage of the polarization and removal of cytoplasmic lipids from the cells before vitrification. The authors thus avoided a negative aftereffect caused by the cooled intracellular lipids. While the intracellular lipids are considered as an energy source for oocytes and as building material for membranes of future embryos, their removal did not adversely affect the development of oocytes and embryos after cryopreservation.

The success of cryopreservation depends on the ability to control, in addition to the cooling rate, the nucleation temperature T_n. T_n depends on the sample volume, properties of the container wall, and composition of the solution. T_n is also found to coincide with the onset of the liquid crystalline to gel phase transition in the cell membranes. Addition to the solution (similar to the approach developed in the freeze-tolerant animals) of ice nucleating substances such as *Pseudomonas Syringae* bacterium or nucleator, known for over half a century, silver iodide AgI shifts T_n to higher temperature but does not allow its precise control. In practice, initiation of crystallization (*seeding*) is achieved manually by touching the sample with cold tweezers or a cold rod. Some years ago, Petersen et al. described a device for controlling the nucleation temperature with a high accuracy for the cryopreservation system containing up to eight samples. The action of the device is based on electrofreezing - an ability of high voltage applied to metal electrodes to cause nucleation in supercooled water, as was demonstrated for the first time by Rau in 1951. Petersen and collegues studied the influence of common cryobiology additives such as glucose, hydroxyethylstarch, glycerol on this effect and showed that it is also observed in the solutions of nonionic additives. However, the presence of ionic NaCl makes electrofreezing in solutions with physiological salt concentration impossible. This problem was solved by the authors by designing a separate volume of pure water inside the cap of platinum electrode that makes nucleus formation to be independent of the solution composition. It should be noted that cryopreservation is not the only field to benefit from the precise control of nucleation temperature - this parameter is known to strongly affect the rate of lyophillization.

Liquid nitrogen could be exposed to either atmospheric pressure or vacuum. The latter decreases the temperature by several degrees of centigrade (at 300 mmHg pressure) down to -200 °C. This improvement in CP was shown to significantly increase the development rate of bovine oocytes after vitrification.

For a long time all attempts at fish embryo conservation have failed. Only in 2007, Robles et al. have reported some success in cryopreservation of two-cell embryos of seabream *Sparus aurata*, using the technique of microinjection of CPA into the embryo to overcome the permeability barrier. The results also support the hypothesis that AFP (the authors used a natural antifreeze protein type I), in addition to inhibition of ice growth and recrystallization, stabilizes the cellular membrane.

There are three ways to increase cooling rates viz.:

1. To minimize the volume of the solution surrounding the cells
2. To minimize the thermo-insulation
3. To avoid liquid nitrogen vapor formation

The simplest way to minimize the thermo-insulation is to remove it by dropping the sample directly into the liquid nitrogen. However, this approach has a number of disadvantages. To form a drop, a rather large amount of the solution is needed. The drop will float for a long period of time on the surface of the liquid nitrogen. A strong evaporation on the surface will form a "vapor coat" as a thermo-insulated layer. One of the approaches to solve this problem is to drop the solution containing sample cells onto the precooled metal plate (solid surface vitrification).

It is known for a long time that the ice nucleation temperature can be reduced by an increase of the hydrostatic pressure, while the glass transition temperature rises with increased pressure. This approach to the reduction of the needed CPA concentration is, however, limited, since high pressure can cause damage to the biological objects. For example, while dogs kidneys survive a 20-min exposure to 1,000 atm, rabitt kidneys were severely damaged after 20 min at 500 atm. It should be note that high pressure is necessary for vitrification only – atmospheric pressure is sufficient for the storage.

Recently, a device for ultra-fast cooling has been suggested by A. Jiao et al. based on the oscillating motion heat pipe (OHP) that provide the extremely high heat exchange coefficient using the thin film evaporation effect in the oscillation motion of the liquid plugs and vapor bubbles. The authors state that the heat transfer coefficient over 10^4 Wm^{-2} K^{-1} could be reached, which will provide, at least for some cooling methods such as ultra-thin straw (100 μm in diameter), cooling rates exceeding 2×10^4 K min^{-1} that allows five-fold reduction of the passage time through the dangerous temperature region (DTR), typically from 240 to 200 K where most ice nucleation occurs, in comparison with conventional cooling techniques and thus significantly reduce the concentration (and, hence, toxicity) of the CPA solutions.

A special procedure has been developed for the CP of blastocysts that contain a fluid-filled cavity called the blastocoele. Since the probability of ice formation is proportional to the sample volume, this cavity is considered a weak point for the CP. In addition, CPA permeation into the blastocoele is slow: the resulting CPA concentration is insufficient after 3-min exposure of blastocysts to EG solution. Vanderzwalmen et al. showed that survival rate could be improved by artificial reduction of the blastocoele with a needle or pipette before vitrification.

CPA solutions

As a rule, there is no distinct difference between cryoprotective solutions for the hypothermic preservation and cryopreservation. Usually any hypothermic preservation solution as well as a buffered physiological solution could be used for cryopreservation. Still, there are exceptions: for example, the so-called St. Thomas Hospital solution for the heart preservation is not good as a long storage medium.

Different additive are aimed at reducing the detrimental effects that accompanies cooling and freezing. Some solutions have been developed specially for the preservation of a particular organ. For example, UW solution was developed first for the liver preservation while the original Collins solution for kidney preservation. The latter tries to mimic the intracellular composition and is rich in K^+. Cardioplegic solutions differ in the basic ionic composition, which can be either intracellular (rich in K^+) or extracellular (rich in Na^+).

Sometimes different solution is used at the preliminary stages of CP: for example, spermatozoa are incubated in the cholesterol-loaded cyclodextrin, since loss of the cholesterol from the plasma membrane is one of the reasons of its destabilization.

A number of CPA mixtures has been suggested aiming primarily at the reduction of the solution toxicity: a low ionic-strength vehicle solution (LSV) containing propanediol and trehalose, Euro-Collins solution, Hank balanced salt solution (HBSS), histidine–lactobionate (HL) solution, histidine–tryptophan–ketoglutanate (HTK) solution, sodium–lactobionate–sucrose (SLS) solution, to name a few.

Sometimes other components called extenders as citrates or egg yolk are added to CPA solution; these compounds are thought to have an additional protective effect during freezing and thawing. The addition of polymers with high molecular weight such as polyvinylpyrrolidone (PVP) or polyethylene glycol (PEG) is aimed at assisting virtification, primarily the extracellular one, since cells contain large number of macromolecules similar in action.

The variety of the cryoconservation procedures and even greater variety of cryoprotective solutions sometimes is rightfully considered as an obstacle to the development of this technique since, in the absence of the evident best approaches, it disperse the research efforts instead of focusing on perfecting a single approach.

A milestone in the development of cryoprotective solutions is the invention in 1986 by Belzer and Southard from the University of Wisconsin of Madison, the synthetic solution called UW-solution that later became the practical standard in cryopreservation – it is usually a starting solution when a new preservation protocol should be developed. The solution contains potassium lactbionate, adenosine, glutamine, allopurinol, hydroxyethyl starch, dexamethasone, and insulin. The comparative study proving its preference over other solutions was reported 2 years later. It was also found that the solution still could be improved by addition of natural factors (small proteins called trophic factors) increasing the storage time, and decreasing the organ damage by stimulation of DNA repair. The recent natural components used in the storage solution are green tea polyphenolic compounds (GTPC). GTPC could act as biological antioxidant and protect mammalian cells and tissues from oxidative stress-induced damage.

Recent study by Wusterman et al. have shown that a significant reduction of the CPA solution toxicity could be attained by replacing sodium in the vehicle solution with choline. In the experiments, twofold and fourfold decrease of the fraction of cells losing their functional capacity were found for the porcine endothelial cells and muscle cells in suspension, respectively. As the authors state, the molecular mechanism by which sodium leads to cell death is still unclear, but it is probably mediated by hydrogen peroxide or nitride oxide. This opinion is supported by the earlier studies of the sodium effects in cardio-myocytes, hepatocytes, and fibroblasts. The advantage of choline chloride over sodium chloride is believed to be its extracellular character with the similar colligative properties that prevents the damaging increase in the intracellular sodium concentration during cryoconservation using the conventional solutions. Beneficial choline properties were also reported for cryopreservation of oocytes and embryos.

In spite of the great progress on cryopreservation and its role in the assisted reproduction (now it is possible, for example, to separate in the course of preservation X- and Y-bearing spermatozoa, thus allowing selective fertilization of oocytes to produce either female or male offspring), there are many unanswered question such as variability of the reaction of spermatozoa from different men to the same cryopreservation procedure or why the fraction of the sperm cells being damaged by freezing is about one half regardless of other conditions.

Cryopreservation is the cornerstone of the rapidly developing branch of medicine - regenerative medicine. Of particular importance is the ability to preserve engineered tissues such as veins, arteries, cartilage and, in perspective, even whole laboratory-produced organs to provide the needed supply for transplantation and allow the necessary gap in time between the moment a replacement is created and its final transplantation. Another fascinating issue is the cryopreservation of human embryonic stem cells that could be lately used for regeneration of certain tissue or given back to the donor to correct age-related deficits (sometimes this emerging discipline is referred to as rejuvenatory medicine).

The difficulties of the whole organ cryopreservation, as was already mentioned, are primarily the different nature of the cells that constitute the organ (and, hence, different optimal preservation procedures) and the significant size that prevents the attaining of the spatial uniformity in both the temperature distribution and tissues saturation with CPAs.

Some progress was reported, but up to 2005 no vital organ (particularly heart or liver) has been deeply frozen to the low enough temperature needed for the long-term storage and later thawed, transplanted, and proved to be functionally consistent. One of the major reasons of failure is the vascular damage due to the mechanical action of ice crystals on the vascular wall. Thus vitrification seems to be the only promising way for the whole organ cryopreservation. Since ultrarapid cooling for large samples is evidently physically impossible, use of highly concentrated vitrification solutions that provide extreme increase of viscosity is inevitable. Recent review of the progress in vitrification of organ can be found in the cited paper by Fahy et al.

Simulation of Solidification

Unfortunately, there is no silver bullet - the single best numerical approach to the study of solidification process. The researcher is forced to make a choice of both the problem formulation and of the numerical method.

Solidification is the phase transition of the first order that involves energy release due to the self-organization of the solid phase - the latent heat of fusion. The boundary between phases that usually extends several molecular layers (liquid near the interface could contain small crystallites and the solid could have a rough surface) could be considered as either a sharp interface where a discontinuity in some of the system variables is present or as a thin transition layer where all thermodynamic parameters vary continuously. Sometimes these two approaches are referred to as the Gibbs approach and the van der Waals approach, respectively. The former is surely more common, especially in the problems with the simple interface geometry.

In the general case, a number of constitutive equations of material are involved in the formulation of the solidification/melting problem. Bene´s has listed these relations, subdividing them into those that relate to the bulk materials and those that relate to the interface between the phases:

1. Constitutive equations of material in the liquid (l) and solid (s) phases:
   (a) Bulk enthalpy per unit volume $H_s(T)$, $H_l(T)$
   (b) Bulk entropy per unit volume $S_s(T)$, $S_l(T)$
   (c) Bulk free energy per unit volume $F_s(T) = H_s(T) - TS_s(T)$, $F_l(T) = H_l(T) - TS_l(T)$
2. Constitutive equations for the interface:
   (a) Interfacial energy per unit area $e(T)$
   (b) Interfacial entropy per unit area $s(T)$
   (c) Interfacial free energy per unit area $f(T) = e(T) - Ts(T)$

The latent heat per unit volume is defined as $L = H_l(T^*) - H_s(T^*)$, where the transition temperature T^* is defined as the temperature at which the free energies of the phases are equal $F_l(T^*) = F_s(T^*)$.

Sharp Interface Methods

The Stefan problem is formulated for the sharp interface approach for the domain Ω by subdividing it by the (unknown) interface Γ into two subdomains – liquid Ω_l and solid Ω_s, so that $\Omega = \Omega_l \cup \Omega_s \cup \Gamma$ and could be stated as follows. The heat conduction equation is to be solved in both the domains:

$$\frac{\partial H_1}{\partial t} = -\nabla q_1 \quad \text{in} \quad \Omega_1(t), \qquad ...(1)$$

$$q_1 = -\lambda_1(T)\nabla T, \qquad ...(2)$$

$$\frac{\partial H_s}{\partial t} = -\nabla q_s \quad \text{in} \quad \Omega_s(t), \qquad ...(3)$$

$$q_s = -\lambda_s(T)\nabla T \qquad ...(4)$$

Here λ is the thermal conductivity and q is the heat flux.

The following conditions should be satisfied at the interface Γ:

$$T = \frac{H_1\big|_1 - H_s\big|_s - k_\Gamma e}{S_1\big|_1 - S_s\big|_s - k_\Gamma s}, \qquad \text{...(5)}$$

$$(q_1 - q_s)n_\Gamma = \upsilon_\Gamma (H_1\big|_1 - H_s\big|_s) - k_\Gamma - D_t e, \qquad \text{...(6)}$$

where n_Γ is a unit vector to $\Gamma(t)$ pointing out of Ω_s, υ_Γ is the normal velocity of the interface, D_t is the derivative with respect to time at $\Gamma(t)$, and $k_\Gamma = \nabla \cdot n_\Gamma$ is the mean curvature of the hypersurface $\Gamma(t)$.

Under an assumption that the enthalpy in the solid and liquid phases is defined as

$$H_s = \int_0^T \rho_s(u)c_s(u)du, \quad H_1 = \int_0^T \rho_1(u)c_1(u)du + L,$$

where $\rho(T)$ and $c(T)$ are the density and the heat capacity of the corresponding phases of the material and denoting the difference in entropy per unit volume as $\Delta s = S_1\big|_1 - S_s\big|_s$, the general formulation of the Stefan problem could be simplified to

$$\rho(T)c(T)\frac{\partial T}{\partial t} = \nabla \cdot (\lambda(T)\nabla T) \quad \text{in} \quad \Omega_1 \text{ and } \Omega_s, \qquad \text{...(7)}$$

$$\lambda_s(T)\frac{\partial T}{\partial n_\Gamma}\bigg|_s - \lambda_1(T)\frac{\partial T}{\partial n_\Gamma}\bigg|_1 = L\upsilon_\Gamma, \qquad \text{...(8)}$$

$$T - T^* = -\frac{\sigma(T)}{\Delta s}k_\Gamma - \alpha\frac{\sigma(T)}{\Delta s}\upsilon_\Gamma. \qquad \text{...(9)}$$

Here σ is the surface tension between the liquid and solid phases and α is the coefficient of the attachment kinetics.

The difference of the temperature at the interface and the transition temperature is usually referred to as kinetic undercooling or the Gibbs–Thomson effect. At the external boundaries of the computational domain, either Dirichlet boundary conditions for the temperature or the Neumann boundary conditions for the heat flux are specified.

The most specific feature of the Stefan problem is the unknown boundary between the liquid and solid domains - crystallization front. Sometimes other moving boundary problems are referred to as Stefan problems. Lame and Clapeyron were the first to consider such problems in 1831; these problems, however, were named after Stefan, who more than half a century later studied the melting of the polar ice cap.

Most of the numerical methods used to solve the Stephan problem are divided into two-domain (three-domain in the case of the explicit treatment of the mushy zone) or front tracking methods in which the interface movement is monitored explicitly, and single-domain (one-domain) methods in which the interface position could be reconstructed as a post-processing procedure. The classification is self-evident; similar division of methods used in the simulation of the gas flows with shock waves using Euler equations for the inviscid fluid, the second group being usually called front-capturing method. Somewhat aside particle and cellular automata methods stand. Numerical methods for tracking discontinuous fronts and interfaces could be subdivided into several groups.

In *surface tracking* methods, the interface is identified by a set of marker points, between which the interface position is approximated by an interpolant. The solution along the interface may be multivalued to account for the discontinuities. The evolution of the interface is usually governed by the differential equations of the lower dimensionality that are derived, as a rule, by the application of Gauss–Ostrogradskii type theorem to the constitutive equations, using the local curvilinear coordinate system orthogonal to the interface. The implementation of these methods consist of two distinct stages: interface updating and interface smoothing. The common problem of the methods of this class is the sensitivity of the interface to the numerical noise.

Volume tracking methods are best for the description of the interaction of the relatively smooth moving interfaces. The interface tracking equations have the same dimensionality as the underlying PDE of the model; however, they need to be solved in the narrow strip containing the interface. The best known member of this family of methods is the famous MAC (marker and cell) method developed over 40 years ago by Harlow and colleagues, another one is the volume of fluid (VOF) method based on monitoring of fraction of each material in every computational cell.

Moving mesh methods include local adjustment methods or Langrangian methods; one of the most flexible variant of the latter is the so-called arbitrary Lagrangian Eulerian (ALE) method that combines advantages of these two approaches to the description of the moving media.

Classical Stefan problem

The classical Stefan problem deals with solidification or melting in the so-called pure substances. Pure substance is an idealization allowing one to disregard diffusion processes - solidification is driven by the temperature evolution alone. Thermomechanical properties are assumed to be constant.

In the absence of fluid convection in both Ω_l and Ω_s subdomains, heat conduction equation is to be solved:

$$\rho c \frac{\partial T}{\partial t} = \lambda \Delta T,$$

coupled by the condition on the interface (sometimes called the Stefan condition) that expresses the continuity of the heat flux, with account for the surface heat source or sink at the interface due to crystallization or melting, value for which is the product of the normal component of the interface velocity υ_γ and the latent heat L of the phase transition:

$$\lambda_s \left.\frac{\partial T}{\partial n_\Gamma}\right|_s - \lambda_l \left.\frac{\partial T}{\partial n_\Gamma}\right|_l = L\upsilon_\Gamma, \qquad \text{...(10)}$$

where T is the temperature, ρ is the density, c is the heat capacity, λ is the thermal conductivity, and n_Γ is the outer normal to the solid subdomain.

Exact solutions

Some analytical solutions have been obtained for the one-dimensional Stefan problem in different coordinate systems. Stefan had obtained an analytic solution to the so-called one-phase problem, assuming the thermal parameters to be constant, and showed that the rate of solidification or melting in the semi-infinite region is governed by a dimensionless parameter

$$St = \frac{C_l(T_l - T_m)}{L},$$

which later was named Stefan number. Here C_l is the heat capacity of the liquid, L is the latent heat of fusion, T_l is the temperature of the surrounding media, and T_m is the melting temperature.

Plane Solidification

John von Neumann obtained a solution to a more realistic two-phase problem. The melting of the semi-infinite region ($0 < x < \infty$) under the uniform initial temperature $T_s \leq T_m$ is considered. The constant temperature is imposed at the boundary $x = 0$, and the thermal properties are assumed to be constant. The following equations constitute the problem formulation:

Heat conduction in the liquid region ($0 < x < X(t)$, $t > 0$):

$$\frac{\partial T_l}{\partial t} = \alpha_l \frac{\partial^2 T_l}{\partial x^2}.$$

Heat transfer in the solid region ($X(t) < x$, $t > 0$):

$$\frac{\partial T_s}{\partial t} = \alpha_s \frac{\partial^2 T_s}{\partial x^2}.$$

The temperature at the interface:

$$T(X(t),t) = T_m.$$

Stefan condition (a balance of the heat fluxes):

$$k_s \frac{\partial T_s}{\partial x} - k_1 \frac{\partial T_1}{\partial x} = L\rho \frac{dX}{dx}.$$

To close the system initial

$$T(x,0) = T_s < T_m$$

and boundary conditions

$$T(0,t) = T_l > T_m,$$
$$T(x,t) = T_s \quad \text{for} \quad x \to \infty,\ t > 0$$

are specified.

von Neumann using the similarity variable $\eta = x/2\sqrt{\alpha_1}$ had obtained an analytical solution that is expressed as follows (erf and erfc are the error function and the complimentary error functions, respectively):

Position of the interface between the solid and liquid domains

$$X(t) = 2\lambda\sqrt{\alpha_1 t}.$$

The temperature in the liquid phase

$$T(x,t) = T_1 - (T_1 - T_m)\frac{\operatorname{erf}(x/2\sqrt{\alpha_1 t})}{\operatorname{erf}(\lambda)}.$$

The temperature in the solid phase

$$T(x,t) = T_s + -(T_m - T_s)\frac{\operatorname{erfc}(x/2\sqrt{\alpha_s t})}{\operatorname{erfc}(\lambda\sqrt{\alpha_1/\alpha_s})}.$$

The parameter λ is determined from the equation

$$\frac{St_1}{\exp(\lambda^2)\operatorname{erf}(\lambda)} - \frac{St_s\sqrt{\alpha_s}}{\sqrt{\alpha_1}\exp(\alpha_1\lambda^2/\alpha_s)\operatorname{erfc}(\lambda\sqrt{\alpha_1/\alpha_s})} = \lambda\sqrt{\pi},$$

where

$$St_1 = \frac{C_1(T_1 - T_m)}{L}, \quad St_s = \frac{C_s(T_m - T_s)}{L}.$$

Cylindrical Solidification

Paterson in 1952 had obtained a similar solution for the case of the cylindrical coordinates. A line heat source of strength Q is located at the symmetry axis of the infinite solid body held under the temperature $T_s < T_m$. The heat source is activated at time $t = 0$. The energy balance near the heat source could be written as

$$\lim_{r\to 0}\left[-2\pi k_1 \frac{\partial T_1}{\partial r}\right] = Q.$$

The Paterson solution is expressed as follows:

Position of the interface

$$R(t) = 2\lambda\sqrt{\alpha_s t}.$$

The temperature in the liquid phase

$$T(r,t) = T_s + \frac{T_m - T_s}{Ei(-\lambda^2\alpha_s/\alpha_1)} Ei\left(-\frac{r^2}{4\alpha_s t}\right).$$

The temperature in the solid phase

$$T(r,t) = T_m - \frac{Q}{4\pi k_s}\left[Ei\left(-\frac{r^2}{4\alpha_s t}\right) - Ei(\lambda^2)\right].$$

The parameter λ is determined from the equation

$$-\frac{Q}{4\pi}e^{-\lambda^2} + \frac{k_1(T_m - T_s)}{Ei(-\lambda^2\alpha_2/\alpha_1)}e^{-\lambda^2\alpha_s/\alpha_1} = \lambda^2\alpha_s\rho L,$$

where Ei(x) is the exponential integral function:

$$Ei(x) = \int_{-x}^{\infty}\frac{e^{-t}}{t}dt.$$

Freezing of Porous Media

The case of crystallization with the explicitly described mushy zone - an extension of the von Neumann solution - was obtained by Lunardini in 1985, who studied the freezing of the porous media. The solution is written as follows:

$$T_1 = (T_m - T_s)\frac{\mathrm{erf}(x/2\sqrt{\alpha_1 t})}{\mathrm{erf}(\psi)} + T_s,$$

$$T_2 = (T_m - T_f)\frac{\mathrm{erf}(x/2\sqrt{\alpha_4 t}) - \mathrm{erf}(\gamma)}{\mathrm{erf}(\gamma) - \mathrm{erf}(\psi\sqrt{\alpha_1/\alpha_4})} + T_s,$$

$$T_3 = (T_0 - T_f)\frac{-\mathrm{erfc}(x/2\sqrt{\alpha_3 t})}{\mathrm{erfc}(\gamma\sqrt{\alpha_4/\alpha_3})} + T_0,$$

where T_0 is the initial temperature, T_m and T_f are the solidus and liquidus temperatures, T_s is the boundary temperature, α_1 and α_3 are the thermal diffusivities of the regions 1 and 3, respectively, defined as

$$\alpha_1 = \frac{k_1}{C_1}, \quad \alpha_3 = \frac{k_3}{C_3},$$

C_1 and C_3 are the volumetric heat capacities of these regions.

The thermal diffusivity of the mushy region is assumed to be constant with latent heat term included

$$\alpha_4 = \frac{k_2}{C_2 + \frac{\gamma_d L \Delta\xi}{(T_f - T_m)}},$$

where $\gamma_d = (1 - \varepsilon)\rho_s$ is the dry unit density of the media solids, $\Delta\xi = \xi_0 - \xi_f$, ξ_0 and ξ_f are the ratio of the unfrozen water to the media solid mass for the fully thawed and frozen states, respectively, given as

$$\xi_0 = \frac{\varepsilon\rho_w}{(1-\varepsilon)\rho_s}, \quad \xi_f = \frac{\varepsilon S_{wres}\rho_w}{(1-\varepsilon)\rho_s}.$$

Here S_{wres} is the residual saturation.

The evolution in time of the position of the boundaries of the mushy and solid regions is given as

$$X_1(t) = 2\psi\sqrt{\alpha_1 t}$$

and

$$X(t) = 2\gamma\sqrt{\alpha_4 t}.$$

The parameters ψ and γ are defined by the following system of equations:

$$\frac{(T_m - T_s)}{(T_m - T_f)} e^{\psi^2(1-\alpha_1/\alpha_4)} = \frac{k_2/k_1\sqrt{\alpha_1/\alpha_4}\,\mathrm{erf}(\psi)}{\mathrm{erf}(\gamma) - \mathrm{erf}(\psi\sqrt{\alpha_1/\alpha_4})},$$

$$\sqrt{\alpha_3/\alpha_4}\frac{(T_m - T_f)k_2/k_3}{(T_0 - T_f)} e^{\gamma^2(1-\alpha_4/\alpha_3)} = \frac{\mathrm{erf}(\gamma) - \mathrm{erf}(\sqrt{\alpha_1/\alpha_4}(\psi)}{\mathrm{erfc}(\gamma\sqrt{\alpha_4/\alpha_3})}.$$

In the classical Stefan problem as well as in the majority of the simulation studies the densities of the liquid and solid phases are assumed to be equal. Generally, it is not so and the difference in density could be significant, for example, in water crystallization. Griebel et al. considered this case. Conservation of mass across the phase boundary allows to equate the normal mass fluxes in both phases at the phase boundary to get

$$\mathrm{v}\cdot\mathrm{n}=\left(1-\frac{\rho_s}{\rho_1}\right)\mathrm{V}_\Gamma\cdot\mathrm{n},$$

where v is the velocity of the liquid phase and V_Γ is the crystallization front velocity.

The temperature at the interface is continuous $T_s = T_l$; to formulate the second condition, the authors considered the total energy balance across the interface that is expresses as

$$[\rho e(\mathrm{v}-\mathrm{V}_\Gamma)+\mathrm{q}\cdot\mathrm{n}-\mathrm{n}^{\mathrm{T}}P\mathrm{n}]_s^1=-\gamma K\mathrm{V}_\Gamma\cdot\mathrm{n},$$

where $e = u + \mathrm{v}\cdot\mathrm{v}/2$, the internal energy is defined in liquid and solid phases as $u_l = c_l(T - T_m)$ and $u_s = c_s(T - T_m) - L$, respectively, and the term in the right hand side related to the interfacial energy originates from the Gibbs–Thompson effect, P is the stress tensor of the Navier–Stokes equations

$$P^{ij}=-p\delta_{ij}+\mu\left(\frac{\partial v_i}{\partial x_j}+\frac{\partial v_j}{\partial x_i}\right)+\lambda\nabla\mathrm{v}\delta_{ij},$$

where μ and λ are viscosities. Details of derivation and the final relations could be found in the cited apper.

Diffuse Interface Methods

The van der Waals approach to the solidification was revived by Cahn and Hillard in the middle of the twentieth century and applied to problems where the microstructure of the forming solid is of interest, for example, for the study of spinodal decomposition. The best known implementation of this approach is the so-called phase field theory methods for the solidification simulation. There are two other classes of problems where the diffuse interface approach seems to be preferable.

Crystal nucleation involves the formation of heterogeneous fluctuations containing crystal-like atomic arrangements. Those fluctuations grow whose size exceeds some critical value determined by the interplay of the interfacial and volumetric contributions to the free energy of the cluster that typically contains tens to hundreds atoms or molecules. As was mentioned above, the classical nucleation theory that assumes a sharp interface between the nucleated embryo and the mother media and uses bulk materials properties for the critical nucleus encounters difficulties in the case of small critical radius. Evidently, the sharp interface approach could hardly provide an adequate description of small clusters whose size is comparable to the physical interface thickness.

The other problem that is difficult to describe using the sharp interface approximation is the polycrystalline growth that usually proceeds in one of the two main modes:

1. Impingement of independently nucleated single crystals (primary nucleation)
2. Nucleation of the new crystalline grains at the perimeter of the existing particles (secondary nucleation)

The most advanced implementation of the diffuse (nonsharp) interface approach is the phase field theory based on introduction of the order parameters that define the local thermodynamic state of the material. The simplest case involves a single order parameter – phase field φ that monitors the interface between the liquid and the solid states. Other parameters (usually this model is referred to as vector-valued phase field model) could be the chemical composition or the orientation field that specifies the orientation of the crystallographic planes of the growing crystal.

The thermodynamic functional F (frequently, the free energy) is expressed via the volumetric density that depends on a set of the order parameters φ_1, $\varphi_2, \ldots,$ φ_N:

$$F = \int_{\Omega} (w(\varphi_1, \varphi_2, \ldots, \varphi_N) + G(D\varphi_1, D\varphi_2, \ldots, D\varphi_N))dx.$$

This functional includes contributions from the bulk density $w(\ldots)$ and a gradient term G that accounts for the nonuniformity of the system. Frequently w is of the multiwell types and so there are some preferable states of the system.

The system evolution in time expressed as the functional relaxation to the minimum value is described by the so-called Model A equation:

$$\tau_i \frac{\partial \varphi_i}{\partial t} = -\delta_{\varphi_i} F[\varphi_1, \varphi_2, \ldots, \varphi_N],$$

where $\delta_{\varphi i} F$ is the Frechet derivative and τ_i is a relaxation parameter.

If the system state is described by single parameter, it could distinguish between liquid and solid or ordered and disordered states of the system. The equation for the evolution of the order parameter in case of crystallization is solved together with the energy equation that governs the heat transfer. The importance of the phase field theory methods is expected to increase in future due to both its ability to cope with complex interfaces encountered such as the branched dendrite structures that are difficult for the sharp interface treatment and possibilities provided by the progress in the computer hardware.

Thermal Properties of Tissues

Model-based optimization algorithms for the planning of the thermal treatment, both hyperthermia and cryosurgery, will give reliable predictions if the thermal properties of tissues are known with a sufficient accuracy. Unfortunately, these properties are poorly known, especially for temperatures below –40 °C, and are often approximated by those of water.

The experimental methods to determine thermal properties of biological tissues are rather advanced for the case of homogeneous samples. The measurements of spatially varying properties are more involved, especially when simultaneously both the thermal conductivity and the heat capacity are to be determined. The additional difficulty of the measurement of the thermal properties of biological tissues in vivo is the need to estimate and separate the blood perfusion contribution to the heat transfer. Finally, the measurement techniques should be noninvasive and be able to operate with small temperature differences acceptable for tissues. As a rule, one should solve the inverse problem and frequently exploit some model of heat transfer in the biological media.

Huttunen et al. recently reviewed approaches to the determination of heterogeneous thermal properties of biological tissue. The authors themselves used ultrasound-induced heating to provide the necessary temperature gradients in tissues and MR imaging the temperature evolution. This technique was used earlier to measure the properties of homogeneous medium by Cheng and Plewes and by Vanne and Hynynen. Huttunen and coworkers considered the target domain that contains several sub-domains in which the tissue parameters, both thermal and acoustic, can be assumed constant. Note that this approach could provide the temperature-independent properties only. To extract the thermal properties from the temperature measurements, the authors used the classical Pennes' equation, neglecting the metabolic heat source. The semi-discrete Galerkine FEM with piecewise linear basis functions were used.

The thermophysical properties of some specific tissue are known to depend on the number of factors such as age, gender, ethnicity, circadian rhytm, and state of thermoregulatory sweating. The state of the blood vessels greatly affect the measured values of the tissue parameters. Thus, according to experiments performed by Parsons on the heat transfer in human skin, the thermal conductivity could vary from 0.2–0.3 Wm^{-1} K^{-1} in the vasoconstricted state to 0.4–0.9 Wm^{-1} K^{-1} in the vasodilated state.

One approach to estimate the thermal properties of the biological tissues is the volume averaging that is frequently used for the heterogeneous media, such as, for example, soil:

$$\lambda = \varepsilon S_w \lambda_w + \varepsilon S_{ice} \lambda_{ice} + (1-\varepsilon)\lambda_s,$$

where ε is the porosity, λ_w, λ_{ice}, and λ_s are the thermal conductivities of liquid water, ice, and solid matrix, respectively, and S_w and S_{ice} are the saturation of liquid water and ice.

The averaging based on the mass fractions of water, proteins, and fats in the biological objects with different weights is common in both medicine and food industry; frequently the coefficients suggested many year ago by Cooper and Trezek are used for thermal conductivity, specific heat, and density.

Human

Thermal properties of some human tissues are presented in Table 1.2.

Table 1.2. Properties of human tissues

Tissue	*Density (10^3 kg m^{-3})*	*Thermal conductivity (Wm^{-1} K^{-1})*	*Specific heat (10^3 J kg^{-1} m^{-3})*
Skin, epidermis	1.2	0.21	3.6
Skin, epidermis		0.21	4.32
Skin, dermis	1.2	0.53	3.8
Skin, dermis		0.53	4.56
Fat	0.85	0.16	2.3
Spleen		0.5394	
Spleen	1.05	0.546	
Muscle	1.05	0.642	3.75
Muscle	1.27	0.53	3.8
Lung		0.4506	
Bone, cancellous	1.7	0.582	1.59
Kidney	1.05	0.54	3.9
Kidney	1.05	0.546	3.74
Liver	1.06	0.57	3.6
Liver		0.5122	
Liver	1.05	0.567	

Animals

Sir Winston, The Fellow of the Royal Society, was right - in many respect pig is the closest to human animal and is preferred model compared to other organisms. Animal models that are currently in use include nonhuman primates, rodents (rats, mice, guinea pigs), large animal models (pig, dog, sheep), the chick, and simple animals, including fish, insects, and round worms. Each model system has strengths and weaknesses, depending on the question being addressed.

Thermal conductivities of some animal tissues, adapted from the data by Holmes, are presented in Table 1.3.

Table 1.3. Properties of animal tissues

Tissue	*Animal*	*Thermal conductivity ($W\ m^{-1}\ K^{-1}$)*
Kidney, whole	Rabbit	0.502
Kidney, cortex	Rabbit	0.465
Kidney, cortex	Dog	0.491
Liver	Rabbit	0.493
Liver	Rat	0.498
Liver	Sheep	0.495
Liver	Dog	0.55
Muscle	Rat	0.505
Muscle	Pig	0.518
Muscle	Cow	0.410
Skin	Giraffe	0.442
Skin	Crocodile, back	0.432
Skin	Crocodile, tail	0.334

Latent Heat

One more parameter characterizing biological solutions should be known to provide the reliability of numerical simulations - latent heat of the phase change that is frequently taken equal to that of pure water. Shepard et al. were probably the first to study the phase change properties of the solutions relevant to cryobiology. Han et al., using DSC thermograms, measured the latent heat of different aqueous mixtures of biological relevance, including sodium chloride and phosphate-buffered saline solution with different chemical additives. The authors have considered glycerol and raffinose as examples of CPAs, and antifreeze protein (type III, molecular weight 6,500), and sodium chloride as a cryosurgical adjuvant. They found that latent heats do not correlate directly (i.e., in a linear fashion) with the water content, but could be correlated with the amount of water that participates in the phase change. The measurements gave the value

303.7 and 233.0 Jg^{-1} for water–ice and eutectic phase change, respectively. The reason for the former being smaller than that for pure water (335 Jg^{-1}) warrants further study. Latent heat of crystallization is found to sharply drop for large concentration of glycerol. The authors also found the disappearance of the eutectic phase transition when even a small amount of glycerol was added, which again was observed after addition of AFP. The eutectic crystallization occurs only under a significant supercooling in contrast to the melting that is in good agreement with the phase diagram temperature.

Similar studies have been performed by Devereddy et al. who base the analysis of the measurements, as the authors themself state, "on the full set of heat and mass transport equations." The authors wrote out for the case of spherical crystals growing in the solution that are all assumed to be identical and grow independently in its own pool of liquid. In fact, only mass transfer equation is solved, since estimates of the similarity parameters showed that heat transfer occurs so rapidly that the temperature could be assumed uniform throughout the system. The authors suggested several explanations for the observed reduction of the latent heat in the considered aqueous solutions in comparison with pure water, including transformation of some amount of water into the unfreezable form due to binding to solutes and possible entropic effects due to the ordering of the water in the presence of solutes prior to the phase transition.

Historical Review

There are two reasons to study a response of various biological objects – cells, tissues, organs – to the action of the physical factors of the different nature (ultrasound, electric or magnetic field, radio-frequency (RF) electromagnetic field, ionizing radiation, laser radiation, extreme temperatures – either high or low):

1. To assess the effects of the natural and technogenic impacts on the living organisms;
2. To use physical fields in medicine for both diagnostics and therapy, greatly enhancing the armamentarium of the modern physician.

Biological effects caused by physical factors are not necessarily pathogenic for human: they may or may not result in adverse health effect. Note that the biological effect could be significant even for the weak action that play a role of a trigger in this case. For instance, the human visual system is known to have an extremely high sensitivity:

the energy of a nerve impulse is by several orders of magnitude larger than the energy of the quantum of light that triggered it; the necessary additional energy is provided by the biology system itself, usually in some way through the metabolic processes. The observed frequency-dependent biological effects of the electromagnetic field are usually attributed to the absorption in the certain bands of the biological molecules.

Both the knowledge obtained and the physical fields themselves could be used in the therapy. For example, understanding of the role that the hydroxonium ion H_3O^+ plays in the processes in the tissues subjected to the radiation allowed to propose the injections of glucose as an aid against the tumor cells. The electric field as well as ultrasound could accelerate the drug delivery. Using the physical field for destruction of the pathological tissue is the noninvasive alternative to the classical resection surgery.

Often a number of different factors act simultaneously. Electroporation is based on the alteration or the destruction of the cell membranes by the electric impulses. The increase in the membrane permeability is associated with formation of the nanoscale defects – pores – in the cell membrane when the strength of the electric field exceeds a certain threshold value. If these pores are permanent and do not reseal (*irreversible electroporation*, IRE), this technique is used for the tissue ablation. The reversible electroporation in which the pores reseal is used for the trans-dermal drug delivery or for the delivery of the normally impermeable drugs, for example, in chemotherapy [663]. Either mode of electroporation is accompanied by the Joule heating of the tissues by the electric current.

RF radiation and ultrasound (including the modern technique of *high intensity focused ultrasound* (HIFU)) are used to destroy the malignant tissues by the high temperatures; however, the electromagnetic field directly affects *deoxyribonucleic acid* (DNA) conformations; *molecular dynamics* (MD) simulations show that the electric field causing a reorientation of the protein's dipole moment results in the changes of the secondary structure of the protein and aids its denaturation. Various nonthermal effects are observed in the intense sonic field such as the secondary flows forming in the large and medium blood vessels and the cavitation due to the presence of gas bubbles in the blood (these bubbles greatly increase the contrast in ultrasonic images). Strong absorption of ultrasound energy by the bubbles leads to their oscillation and finally to instability – inertial cavitation.

Both the mechanical stress on the epithelian cells covering the vessels' inner surface caused by the shear stress in the blood flow (that is forming in the viscous fluid flow near the wall or due to the microstreaming surrounding bubbles) and an activation of the processes in the cells at the gene level occur. Laser interstitial thermo therapy (LITT) is aimed at the tissue destruction by the thermal coagulation, but simultaneously photochemical processes at the molecular level could occur; if pulsed lasers are used, mechanical effects (*photoacoustic*) are also taking place.

Low Temperatures in Nature

The normal temperature of warm-blooded (*tachymetabolic*) organisms (and even of some cold-blooded (*bradymetabolic*) ones due to the efficient behavioral thermoregulation through the internal means such as muscle shivering, fat burning, and panting allowing to raise the animal's metabolic rate to produce heat) is in the range 35–42 °C; a stability of the temperature of the human brain is about one-tenth of a degree Centigrade.

Some organisms dye from cooling to moderate temperatures before the freezing occurs. For a human patient being held at 15 °C for the open-heart surgery, a maximal duration of the operation is about 2 h. The accidental profound hypothermia is frequently lethal: the reported mortality related to the progressive reduction of cardiac output and a sudden fall in arterial blood pressure ("*rewarming shock*") is 52–80%, depending on the rewarming method. Low temperatures decrease myofilamental Ca^{2+} sensitivity and increase myocardial contractility; significant changes are already observed for the moderate temperature decrease down to 30 °C.

Numerous types of cells and organisms, however, satisfactorily endure the cooling to the temperatures higher than the crystallization temperature (e.g., the human erythrocytes – up to one and a half months). Many animals including both invertebrate (numerous terrestrial insects, several species of intertidal marine mollusks) and vertebrate (some species of terrestrially hibernating amphibians and reptiles – several species of frogs such as the wood frog *Rana sylvatica*, the gray tree frog *Hyla versicolor*, spring peeper *Pseudacris crucifer*, the box turtles *Terrapene carolina* and *Terrapene orusta*, the Siberian salamander *Salamander keyserlingi*, and the European common lizard *Lacerta vivipara*) endure a partial crystallization of the body water.

Two main strategies for the adaptation of the organism to low temperatures observed in nature lead to the classification of organisms into two groups - freeze-tolerant and freeze-avoiding. The biological mechanisms providing the survival of different species include, inter alia, accumulation of *antifreeze proteins* (AFPs), *ice-nucleating agents* (INA), polyols (glycerol, sorbtol, thereitol, erythritol etc.), sugars (such as fructose, sucrose, and trehalose), and by redistribution of water in the organism. The death of the cells at low temperatures is related, as a rule, to either the protein denaturation or the modification of the cellular membrane structure.

Some animals restore all the functions after freezing of about 50% (according to Layne and Lee — up to 70%) of the total body water (e.g., the wood frog *Rana sylvatica* — 65% for up to 2 weeks; the conversion of up to 80% of body water into ice was reported for barnacles); there are cells (such as isolated fat body cells from insects, e.g., from the freeze-tolerant larva *Eurosta solidaginis*) and organisms that survive the formation of the intracellular ice (the Antarctic nematode *Panagrolaimus davidi*, Pacific oyster *Crassotrea gigas*), the insects survive the temperatures as low as -80 °C, some woody plants down to -50 °C, twigs of woody plants to liquid nitrogen temperature, the invertebrates and the dried seeds down to the temperature close to the absolute zero. J.P. Sparks et al. used time domain reflectometry (TDR) to measure the liquid water content of otherwise frozen material. The authors considered different types of wood (*Robina pseudoacacia*, *Populus trichocarpa*, *Pinus contoria*, *Larix occidentalis*) and found that more than 25% of the water in wood of all species was liquid at temperature from -15 to -50 °C. There are numerous examples of bacteria being revived after long-term storage in the Siberian and Antarctic permafrost.

Sometimes the cold detection is performed by the special cold receptors (such as the so-called bulbs of Kraus residing in the dermis of the skin of humans). In nature, however, far more important is the cellular level of the detection of the low temperatures by organisms and the transduction of the signals in order to activate the biochemical changes that provide defense against cold. The plasma membrane frequently is considered as the principal sensor of low temperatures and the membrane rigidification causes cytoskeleton rearrangement and changes in the ions influx to the cell. Studies of the pathways of the "cold" signal transduction in *Arabidopis thaliana* showed that intracellular calcium plays a significant role in the process: all cell types respond to cold by elevating calcium level; in fact, the elevation of calcium could be caused by different abiotic stresses, including salinity. Symptoms of frost injury in plants are similar to those caused by desiccation. The increase of the intracellular concentration of calcium, in turn, activates the expression of the transcription factors that control the subsequent transcription of several cold-regulated genes. The effects of divalent cations of magnium and calcium on freezing tolerance are similar, but Ca^{2+} is five times more effective than Mg^{2+}.

Cryomedicine: Cryosurgery and Cryopreservation

Low temperatures are used to liquidate the pathological tissues (cryosurgery; also cryodestruction, cryoablation) as well as to suppress the metabolic and other biochemical processes in cells (including stem cells such as human post-natal stem cells needed for stem-cell-mediated clinical therapies and tissue engineering), tissues (natural and composite tissue transplants) and engineered tissues (such as tissue-engineered blood vessels (TEBV) or osteoblast-seeded hydroxyapatite implants as cortical bone substitutes) or organs with a subsequent revitalization for the transplantation, preservation of female fertility (in vitro fertilization (IVF) - preservation of human oocytes (including ones at pronuclear stage with still separate female and male genomes), zygotes, early cleavage-stage embryos, and blastocytes), testing of new pharmaceutical products and the conservation of the genetic resources - both domestic livestock and the rare and endangered species - as well as for facilitating the distribution of the genetically superior domestic species. Cryopreservation of living bacteria is a way to stabilize cultures for the long-term storage and exclude or significantly reduce contamination, phenotypic and genetic drifts, and enable cultures used as standards.

Low temperature preservation usually is subdivided into hypothermic preservation at temperatures above freezing, which is mostly used for the short-term preservation of whole organs and cryopreservation. Two main modes of cryopreservation are the slow controlled freezing and vitrification direct transition to the glassy state without crystallization. Vitrification is based on the notion that the homogeneous nucleation temperature is not a thermodynamic constraint, just a kinetic one and thus is not an absolute but a practical limit depending on the cooling rate.

The biological objects are routinely stored at the liquid-nitrogen temperature; therefore, the only source of damage will be the slow accumulation of the physical damage of DNA from the background ionizing radiation. Whittingham et al. showed that the mouse embryos survive 2 years exposure to the ionizing radiation which intensity is 100-times of the background radiation. Sometimes higher temperatures are used for storage. At temperatures about -80 °C, viabilities decline over weeks to months, while at temperatures below -140 °C, no degradation is observed. The bovine spermatozoa that have been stored for over 50 years show no reduction in viability. The most reliable data is probably on the long-term storage of the human red cells. The study of the samples stored for the period from 11 to 34 years gives the estimate of the shelf life for the conserved human erythrocytes as 113 and 276 years for the storage temperatures -65 °C and -196 °C, respectively.

The term "*cryosurgery*" is not exact in its second part since "hand work" is limited to the preliminary stage of the operation - placing of the cryoprobes. Wikipedia considers "*cryosurgery*" and "*cryotherapy*" as the synonyms (sometimes they are distinguished by the minimal temperature - whether it is higher or lower than the crystallization temperature).

Cryosurgery has a number of advantages over the alternative techniques such as relatively painless and bloodless treatment, possible treatment of outpatients, ease of performance and low cost, high cure rate, few post-surgical complications, short recovery period. Sometimes cryoablation is preferable over the RF ablation even when the local recurrences after these treatment techniques are equivalent due to the greater risk of insufficient ablation by RF in case of metastases located near to the major blood vessels.

The drawbacks of cryosurgery are the short history and scarce information on the long-term efficiency, bulky equipment, expensive MRI or other monitoring, and difficulty in the precise temperature control. In dermatology cryosurgery is used for removal of diverse benign and malignant lessons such as actinic keratosis, warts, hypertrophic scars, keloids, and mucoceles solar lentigos. Cryosurgery treatment of the surface tissues is performed by spraying the liquid nitrogen; the simplicity of this procedure makes it accessible for a family physician. Low temperatures are used in cryoplasty - a procedure to open arteries narrowed due to the atherosclerosis by exploiting liquid nitrous oxide to simultaneously inflate and cool dilatation balloon.

Cryo-insults are used as a research vehicle in animal models as pioneered by Openchowski, with freezing portions of the cerebral cortex of dogs almost two centuries ago. Now they are applied to get information on the early stages of pathogenesis of osteonecrosis, since the resulting repair responses are similar to those for naturally occurring lesions. For instance, K.L. Reed et al. followed this approach to study the femoral head osteonecrosis, which is an important problem in the orthopaedic surgery of the hip using a novel animal model - emu - in which cryosurgically induced necrotic lesions consistently progress to the human-like structural collapse.

Cryoaction could be combined with the traditional therapeutic procedures - adjuvant cryosurgery (e.g., with the immune therapy - or the chemotherapy - chemo-cryo combination therapy) - as well as with other physical methods of the tissues' treatment. The tissue fate in the peripheral zone of the ice ball is critical to the success of the treatment because of the possibility that some tumor cells may survive, thus adjunctive therapy is beneficial. Freezing and chemotherapy differentially activate apoptotic cascades and thus act in parallel. The ^{125}Iodine seed implantation can destroy the residual surviving cancer cells after cryosurgery.

The use of the heater probes or the laser radiation enhances the possibility to control the extent of the frozen domain; for the same purpose injections of the solutions with the special thermophysical properties (e.g., the mixtures of nanoparticles of metal (such as gold) or metal oxide (such as iron oxide magnetite Fe_3O_4) with water that are used in the RF-magnetic hyperthermia) could be used; spraying of the cryogenic fluids is exploited for the protection of the epidermis during the laser operations on the surface tissues; the application of the impulse lasers in cryoconservation to increase the cooling rate is proposed by Kandra. An aggressive pre-cooling could prevent the thermal tissue damage associated with Joule heating in the skin electroporation (especially injury to the highly electrically resistive stratum corneum and epidermis).

The RF heating during the thawing of the cryoconserved biological objects increases the uniformity of the temperature field and prevents the ice recrystallization; RF radiation could also be used for the action at the molecular level. Breaking of the hydrogen bonds inhibits the ice crystal nucleation, thus reducing the crystallization temperature; this mechanism is also considered as a way to reduce the amount of the bound water in the cell and thus to increase the thermal conductivity of the tissue that will enhance the cryodestruction. The considerable (2–3 times) enlargement of the zone of necrosis in the presence of the RF radiation was also observed by Hines-Peralta et al. Cryosurgery is also used to complement dissections and resections, for example, in treating of cardiac arrhythmias. In this case cryosurgery is preferable over the alternative RF ablation, being less thrombogenic. Moderate low temperatures are also used in therapy, for example, the intracarotid cold saline infusion in the human brain to induce the local hypothermia (gray matter is cooling twice as fast as white matter) having advantages over the hypothermia that is induced by the surface cooling in much shorter time to action and lesser probability of such adverse effects as impaired immune function, decreased cardiac output, pneumonia, and cardiac arrhythmias/bradycardias.

Lyophillization

Freeze drying is used in pharmaceutical industry to prepare drugs that endure a long storage time. This process is also used to create the artificial tissues: the porous structure of the collagen sponge – a substitute for a skin – is formed during the lyophillization of the frozen suspension of the collagen protein and is determined by the ice dendrites' morphology. This approach is also investigated for the preservation of (mouse) spermatozoa and human red blood cells.

Cryofixation

Fast freezing is used (sometimes under high pressure) to study biological tissues with a cryomicroscope ("*freezing-shearing*" or "*freezing-etching*"). The key problem of an electron tomography is to immobilize the object under study for the measurement's duration: placing the object into the amorphous ice provides the absence of changes in neither conformations nor relative location of macromolecules and cellular organelles in space.

Study of the structure of proteins and other macromolecules by methods of the neutron crystallography or the X-ray analysis using synchrotron sources requires a preliminary crystallization of the object; the crystal quality including the mosaicity level determines the resolution of the electron density map.

It is implicitly assumed that conformations of the macromolecules in the solution when they are biologically active and in the crystal are identical since "some water molecules are incorporated into the crystal during the growth" and "the crystal is placed into the water during measurements." Indeed, a significant number of the water molecules (from a half to two per every amino acid residue, or about 200 for a typical protein, according to other estimates from 34% to 39% by mass) is incorporated into the growing protein crystal. However, some of these are incorporated into the structures not intrinsic for the protein itself, but formed during the crystal growth. The following should also be noted:

1. Some complexes (such as lipid-protein) are extremely difficult to crystallize;
2. Up to 30–40% of the area of the macromolecule surface that is in contact with the solvent under normal conditions (SAS - solvent accessible surface) is hidden inside the crystal in the case of small proteins;
3. No cold denaturation is observed for the proteins incorporated into the crystal, presumably due to the stabilizing action of the van der Waals forces and the hydrogen bonds between molecules;
4. The crystallization of proteins from the solution requires a supersaturation to be extremely high, orders of magnitude greater than that is usual for the growth of the inorganic crystals; some proteins forms aggregates since, in contrast to the nucleation, there is no energetic barrier to this process and these aggregates act as "the growth units" (e.g., the white hen egg lysozyme forms dimers, tetramers, octamers, and 16-mers) while even formation of the simplest aggregate - the dimer - leads to modifications of the protein structure; in particularity, the hydrophobic groups are shifted to the protein surface.

Another advantage of the cryoelectron microscopy is its applicability to the study of not only macromolecules but also of the heterogeneous structures, for example, subcellular (and even subnuclear) objects in their mutual disposition corresponding to a living cell. The rate of the freezing is critical: Jeffree et al. using a number of plant samples showed that otherwise a redistribution of the water in the tissues occurs, resulting in an appearance of ice deposits in the extracellular space. Sometimes the artifacts are more difficult to detect as, for example, the six- and seven-member ring structures in the hydration shell of the protein that exist at the temperatures below 200 K.

Rapid freezing preserves the structure of proteins to atomic resolution. Electron cryomicroscopic tomography (Cryo ET), however, encounters two difficulties of its own. First, the image contrast is low reflecting small differences in density (more strictly, in electron scattering) between the clusters of macromolecules and cytosol. Second, the samples are sensitive to the damage done by the beam electrons.

Frozen tissues are also used for macroscopic studies: for example, E.M. Brey et al. described the quantitative three-dimensional analysis of microvascular structure that allows to determine such parameters of the system of blood vessels as tortuosity and percentage of endothelial cell area.

Destruction of Biological Tissues

Under the thermal action in contrast to the radiation one, the response of the living tissues is determined by the thermal history: the extent of the domain of action changes due to the heat conduction and the heat transfer by the blood flow. The behavior of tissues in the cryosurgery is more complex than during the hyperthermia treatment, since

1. The ice crystals are formed and grow
2. The ice recrystallization and the formation of gas bubbles can occur during thawing
3. Concentration of the solutes in the cytoplasm and in the extracellular space changes with time
4. The mechanical stresses arising due to both the nonuniformity of the temperature distribution and the volumetric expansion at the phase change can induce fractures in the frozen tissues, causing, inter alia, postoperative bleeding.

The response to freezing injury varies from inflammatory to destructive, depending on the severity of freezing. There are three stages of the living tissue's destruction in cryosurgery.

1. First of all, it is the immediate "direct cell" injury - destruction of the cell due to ice crystal formation and mechanical forces acting on the cell membrane and intracellular ice blades chopping the cell organelles, especially mitochondria, the crystals' size enhancement due to the recrystallization - growth of large crystals at the expense of smaller ones - during thawing (the free energy of the crystal surface decreases with its curvature) or due to the cell solute concentration growth ("solution effect" - exposure of cells to high solute concentrations changes the hydrophobic and hydrophilic interactions and thus functional properties of the membrane lipids, proteins, and nucleic acids). Another mechanism of the cell damage during thawing ("osmotic shock") is related to the decrease of the osmolarity of the surrounding solution due to the ice melting, leading to the shift of free water into the cell and swelling of the latter leading to the cell membrane rapture.
2. Second, a blood microcirculation failure is observed in several hours or days after the operation. Vascular stasis and resulting cellular anoxia are developed through a number of the successive changes in the blood circulation. Initially freezing of tissues causes a vasoconstriction that lead to the decrease of blood flow. The process could include such phenomena as damage of the vascular endothelium, resulting in the increased permeability of the capillary

wall, aggregation of platelets, and formation of microtrombs. Recovering of the circulation during thawing could be accompanied by compensatory vasodilatation and hyperperfusion of the tissues that is thought to stimulate free radical formation, which causes additional endothelial damage via the peroxidation of the membrane lipids.

3. Finally, it is apoptosis (gene regulated cell death) caused by low temperatures that is essential for the partially damaged tissues in the peripheral regions of the ice ball, where the tissue's temperature could be not low enough for the direct ablation during operation. The cells are susceptible to entering the apoptotic state up to 8 h after rewarming. Consideration of the apoptosis as well as of the immune reaction of the organism to low temperatures is beyond the present work.

Frostbite (accidental cryosurgery) is characterized by formation of ice crystals inside the organism under low temperature environmental conditions. It is almost always occurs for the outer extremities, the feet and hands, and face. Frostbite often occurs in conjunction with hypothermia. IIF does not usually occur in frostbite, except perhaps for the skin cells in contact with cold metal. Unlike cryosurgery, which is frequently performed in close vicinity to the vital organ, frostbite does damage important tissues. Necrosis is observed within several days of the injury. The severe frostbite kills nerve cells, so no great pain is associated with it (however, mild frostbite could be painful).

The damage to the cells and tissues from freezing by different mechanisms is cumulative.

An important issue for both cryosurgery and cryopreservation (as well as for hyperthermia) is rapid and reliable characterization of the damage to the biological objects. Different approaches are used, both general ones such as morphological and physiological observation, NMR spectroscopy, electron spin resonance spectroscopy, oxygen or glucose consumption analysis, and cryobiology-specific such as freezing curve-based monitoring or dynamic low-frequency electrical impedance detection of damaged tissues.

Evidently, the conditions causing the tissue's destruction should be avoided in cryopreservation.

Forensic Medicine

The osmotic effect and the crystallization of ice cause the irreversible changes in the tissue: the destroyed cellular bonds are not restored after thawing, that results in the change of the microstructure of the tissue as well as of the mechanical and thermophysical properties; voids appear originating from ice cavities that arise either due to cell membrane fracture or by pushing cells apart, which in some cases (e.g., in the tissues of liver) form regular structures. It is proposed by A.Th. Schafer and J.D. Kauffmann to use the latter fact to inquire whether the tissues were exposed to freezing in the past).

Food Industry

The irreversible character of the changes in the frozen biological tissues - the cellular architecture is not recovered after thawing - is of concern also for the food industry: although freezing change the food products less than other preservation methods, "thaw drip" from meat and fish and the mushy texture of many fruits and vegetables after thawing are observed. The main physical changes in the frozen foods during storage are freeze-cracking, moisture migration, ice recrystallization, and drip losses during thawing.

The change of the palatability characteristics of meat due to the freezing could be assessed quantitatively using such a standard measure of the meat tenderness as Warner–Bratzler shear force. The quality of fruits and vegetables after frozen storage also could be assessed objectively. There are three groups of tests routinely used for this purpose: (1) "fundamental" tests consist in usual engineering measurements of some physical properties, mostly mechanical, such as Young modulus, shear modulus, Poisson ratio, etc.; (2) empirical tests (puncture, compression, extrusion, and others) whose outcomes do correlate with sensory judgement, but our knowledge is insufficient to understand these relations; (3) imitative tests - those that utilize mechanical tools to mimic what occurs in the mouth when the food is masticated. The recrystallization of ice during storage also negatively affects foods that are eaten while frozen such as ice cream.

Two main aids against the damaging effect of freezing on food products - variations of the cooling–thawing protocols (variable-rate freezing) and permeating of the tissues with the edible small molecules - are considered. Among the latter, various polyols including the nonreducing sugars such as raffinose and stachyose are the most common. This practice is based on the observation of the cold-adapted organisms in nature, where the local concentration of sugars could be very high (up to 60% wt:wt). Use of cryoprotectant, however, rises the question whether these chemicals adversely influence taste, texture, or toxicity of frozen foods.

The quality of the frozen food also depends on the cooling rate: inside samples the freezing is slower and the grown crystals are larger. Weight loss during freezing due to dehydration (which is, e.g., around 1.5–2.3% for beef carcass) is economically significant.

Other substances that attract a considerable attention in the food technology research are amino acids that contribute to freezing point depression and to osmoregulation. Rudolph et al. assert that proline has the best membrane-stabilizing effect among the free amino acids; however, this opinion is not shared by other researches. The drip reduction is reported by preliminary meat soaking in the solution of the antifreeze proteins. Antifreeze proteins are very attractive for food technology, since they demonstrate their activity at extremely low concentrations (about 25 μg of protein per liter) and thus unlikely to have any noticeable effect on either taste or texture. To make this approach practical, the synthetic analogs of the antifreeze proteins and glycoproteins found in the polar fishes and the insects should be produced. Recently, the high thermal hysteresis activity of the synthetic analog of AFGP consisting of repeating tripeptide units $(Ala - Thr - Ala)_n$ with a monosaccaride moiety (GalNacl-) attached to each treonyl residue was demonstrated for islets isolated from male Wistar rats. Synthetic AFGP (syAFGP) alters the morphology of the ice crystal into a hexagonal bipyramid, similar to the natural fish-derived AFGP. AFPs are also used to control the crystal size and shape in ice-cream.

Ice nucleating agents could be used in the food industry for the accelerated freezing of foods with the energy savings due to the removal of the excessive supercooling and for facilitating of the lyophillization.

Numerical Simulation of Cryoaction

If Lord Kelvin will be alive today, he probably will continue: "... and can model it." The aims of numerical simulation are to explain (perhaps, using the information that could not be obtained experimentally) and to predict. The latter assumes that the numerical models have been verified (the correctness of the program implementation of the mathematical model has been proved) and validated (the adequacy of the model for the description of the real world phenomena has been proved).

There is, however, a danger related to numerical simulations that now easily produce huge datasets. Half a century ago, Richard W. Hamming in his classical monograph "Numerical Methods for Scientists and Engineers" warned that aim of computations is understanding, not numbers. It is not hard to escape drowning in the ocean of numbers:

one just should remember that more data does not necessary mean more information, more information does not mean more knowledge, and more knowledge does not mean more wisdom.

Cryoaction cardinally differs from the hyperthermic action (ultasound, laser, or RF radiation heating). The latter is based on the *volume* absorption of the acoustic or electromagnetic signal; the cooling of the tissues is possible through the *surface* only. Thus, spatial nonuniformity of the temperature and the rate of its variation, that is, different thermal history of the cells in cryosurgery and cryopreservation is inevitable. In numerical simulation of the hyperthermia, a source term arises in the energy equation; description of the cryoaction is contained in the boundary conditions.

The book contains the description of the physical (in the wide sense, i.e., considering chemistry as a part of the molecular physics) phenomena that occur during freezing and thawing of the biological objects and mathematical models, including numerical ones, used to study these phenomena.

2

WATER

Water (moisture) is the predominant constituent in many foods. As a medium water supports chemical reactions, and it is a direct reactant in hydrolytic processes. Therefore, removal of water from food or binding it by increasing the concentration of common salt or sugar retards many reactions and inhibits the growth of microorganisms, thus improving the shelf lives of a number of foods. Through physical interaction with proteins, polysaccharides, lipids and salts, water contributes significantly to the texture of food.

Table 2.1. Moisture content of some foods

Food	*Moisture content (weight %)*
Meat	65–75
Milk	87
Fruits, vegetables	70–90
Bread	35
Honey	20
Butter, margarine	16–18
Cereal flour	12–14
Coffee beans, roasted	5
Milk powder	4
Edible oil	0

The function of water is better understood when its structure and its state in a food system are clarified. Special aspects of binding of water by individual food constituents and meat are discussed later.

Structure

Water Molecule

The six valence electrons of oxygen in a water molecule are hybridized to four sp^3 orbitals that are elongated to the corners of a somewhat deformed, imaginary tetrahedron. Two hybrid orbitals form O–H covalent bonds with a bond angle of 105° for H–O–H, whereas the other 2 orbitals hold the nonbonding electron pairs (n-electrons). The O–H covalent bonds, due to the highly electronegative oxygen, have a partial (40%) ionic character.

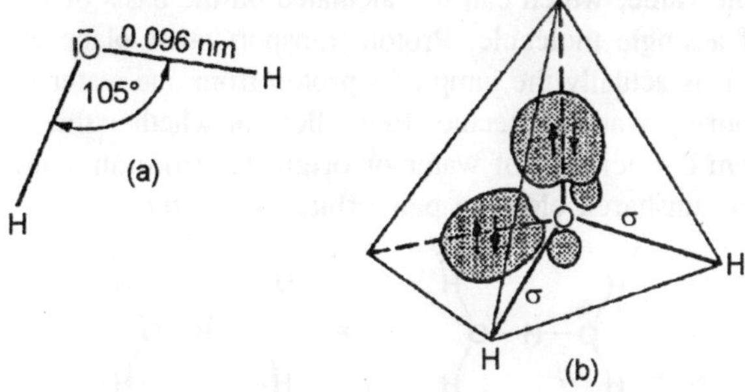

Fig. 2.1. Water. (a) Molecular geometry, (b) orbital model.

Each water molecule is tetrahedrally coordinated with four other water molecules through hydrogen bonds. The two unshared electron pairs (n-electrons or sp^3 orbitals) of oxygen act as H-bond acceptor sites and the H–O bonding orbitals act as hydrogen bond donor sites. The dissociation energy of this hydrogen bond is about 25 kJ $mole^{-1}$.

The simultaneous presence of two acceptor sites and two donor sites in water permits association in a three-dimensional network stabilized by H-bridges. This structure which explains the special physical properties of water is unusual for other small molecules. For

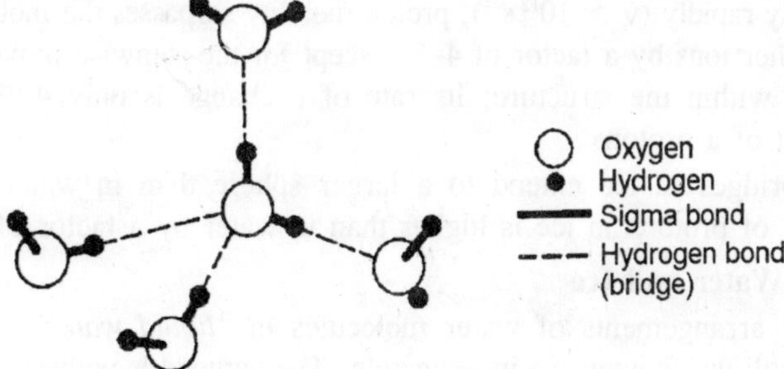

Fig. 2.2. Tetrahedral coordination of water molecules.

example, alcohols and compounds with iso-electric dipoles similar to those of water, such as HF or NH_3, form only linear or two-dimensional associations.

The above mentioned polarization of H–O bonds is transferred via hydrogen bonds and extends over several bonds. Therefore, the dipole moment of a complex consisting of increasing numbers of water molecules (multi-molecular dipole) is higher as more molecules become associated and is certainly much higher than the dipole moment of a single molecule. Thus, the dielectric constant of water is high and surpasses the value, which can be calculated on the basis of the dipole moment of a single molecule. Proton transport takes place along the H-bridges. It is actually the jump of a proton from one water molecule to a neighboring water molecule. Regardless of whether the proton is derived from dissociation of water or originates from an acid, it will sink into the unshared electron pair orbitals of water:

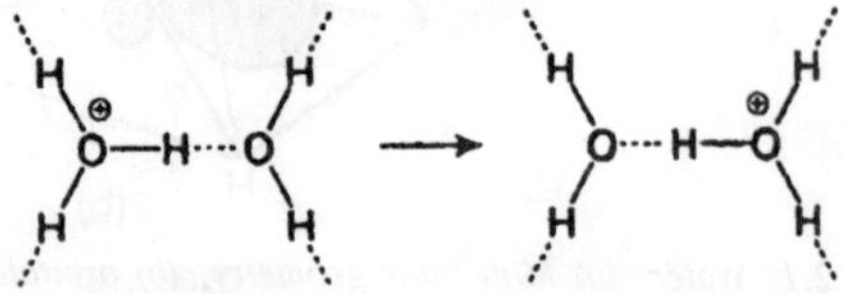

In this way a hydrated $H_3O^{\oplus}$ ion is formed with an exceptionally strong hydrogen bond (dissociation energy about 100 kJ mol^{-1}). A similar mechanism is valid in transport of $OH^{\ominus}$ ions, which also occurs along the hydrogen bridges:

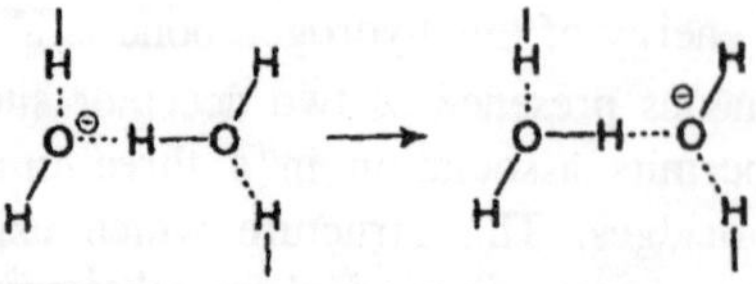

Since the transition of a proton from one oxygen to the next occurs extremely rapidly ($\nu > 10^{12}\ s^{-1}$), proton mobility surpasses the mobilities of all other ions by a factor of 4–5, except for the stepwise movement of $OH^{\ominus}$ within the structure; its rate of exchange is only 40% less than that of a proton.

H-bridges in ice extend to a larger sphere than in water. The mobility of protons in ice is higher than in water by a factor of 100.

Liquid Water and Ice

The arrangements of water molecules in "*liquid water*" and in ice are still under intensive investigation. The outlined hypotheses agree with existing data and are generally accepted.

Due to the pronounced tendency of water molecules to associate through H-bridges, liquid water and ice are highly structured. They differ in the distance between molecules, coordination number and time-range order (duration of stability). Stable ice-I is formed at 0 °C and 1 atm pressure. It is one of nine known crystalline polymorphic structures, each of which is stable in a certain temperature and pressure range. The coordination number in ice-I is four, the O–H···O (nearest neighbor) distance is 0.276 nm (0 °C) and the H-atom between neighboring oxygens is 0.101 nm from the oxygen to which it is bound covalently and 0.175 nm from the oxygen to which it is bound by a hydrogen bridge. Five water molecules, forming a tetrahedron, are loosely packed and kept together mostly through H-bridges.

Table 2.2. Coordination number and distance between two water molecules

	Coordination number	*O–H···O distance*
Ice(0 °C)	4	0.276 nm
Water (1.5 °C)	4.4	0.290 nm
Water (83 °C)	4.9	0.305 nm

When ice melts and the resultant water is heated, both the coordination number and the distance between the nearest neighbors increase. These changes have opposite influences on the density. An increase in coordination number (i.e. the number of water molecules arranged in an orderly fashion around each water molecule) increases the density, whereas an increase in distance between nearest neighbors decreases the density. The effect of increasing coordination number is predominant during a temperature increase from 0 to 4 °C. As a consequence, water has an unusual property: its density in the liquid state at 0 °C (0.9998 g cm^{-3}) is higher than in the solid state (ice-I, ρ = 0.9168 g cm^{-3}). Water is a structured liquid with a short time-range order. The water molecules, through H-bridges, form short-lived polygonal structures which are rapidly cleaved and then reestablished giving a dynamic equilibrium. Such fluctuations explain the lower viscosity of water, which otherwise could not be explained if H-bridges were rigid.

The hydrogen-bound water structure is changed by solubilization of salts or molecules with polar and/or hydrophobic groups. In salt solutions the n-electrons occupy the free orbitals of the cations, forming "*aqua complexes*". Other water molecules then coordinate through H-bridges, forming a hydration shell around the cation and disrupting the natural structure of water. Hydration shells are formed by anions through ion-dipole interaction and by polar groups through dipole-dipole interaction or H-bridges, again contributing to the disruption of the structured state of water.

Aliphatic groups which can fix the water molecules by dispersion forces are no less disruptive. A minimum of free enthalpy will be attained when an ice-like water structure is arranged around a hydrophobic group (tetrahedral-four-coordination). Such ice-like hydration shells around aliphatic groups contribute, for example, to stabilization of a protein, helping the protein to acquire its most thermodynamically favorable conformation in water.

The highly structured, three-dimensional hydrogen bonding state of ice and water is reflected in many of their unusual properties.

Additional energy is required to break the structured state. This accounts for water having substantially higher melting and boiling points and heats of fusion and vaporization than methanol or dimethyl ether. Methanol has only one hydrogen donor site, while dimethyl ether has none but does have a hydrogen bond acceptor site; neither is sufficient to form a structured network as found in water.

Table 2.3. Some physical constants of water, methanol and dimethyl ether

	$F_p(°C)$	$K_p(°C)$
H_2O	0.0	100.0
CH_3OH	–98	64.7
CH_3OCH_3	–138	–23

Effect on Storage Life

Drying and/or storage at low temperatures are among the oldest methods for the preservation of food with high water contents. Modern food technology tries to optimize these methods. A product should be dried and/or frozen only long enough to ensure wholesome quality for a certain period of time.

Naturally, drying and/or freezing must be optimized for each product individually. It is therefore necessary to know the effect of water on storage life before suitable conditions can be selected.

Water Activity

In 1952, Scott came to the conclusion that the storage quality of food does not depend on the water content, but on water activity (a_w), which is defined as follows:

$$a_w = P/P0 = ERH/100$$

P = partial vapor pressure of food moisture at temperature T

P_0 = saturation vapor pressure of pure water at T

ERH = equilibrium relative humidity at T.

The relationship between water content and water activity is indicated by the sorption isotherm of a food.

At a low water content (<50%), even minor changes in this parameter lead to major changes in water activity. For that reason,

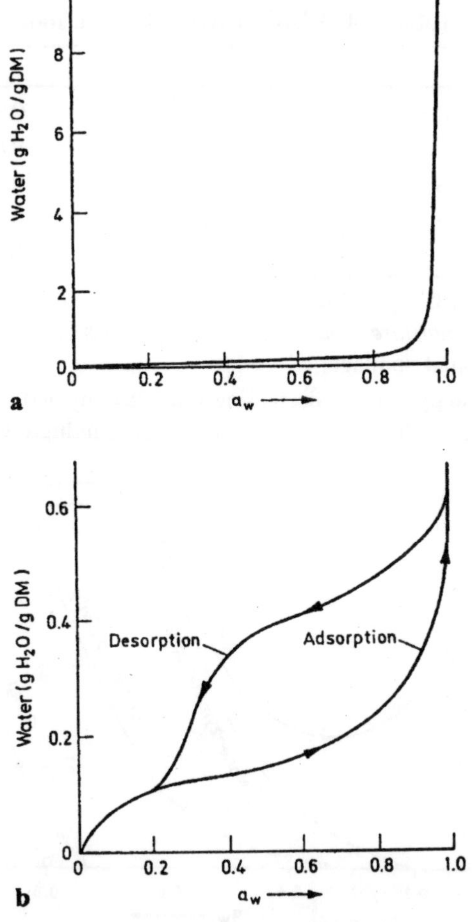

Fig. 2.3. Moisture sorption isotherm. (a) Food with high moisture content; (b) food with low moisture content.

the sorption isotherm of a food with lower water content is shown with an expanded ordinate in Fig. 2.3b, as compared with Fig. 2.3a.

Figure 2.3b shows that the desorption isotherm, indicating the course of a drying process, lies slightly above the adsorption isotherm pertaining to the storage of moisture-sensitive food. As a rule, the position of the hysteresis loop changes when adsorption and desorption are repeated with the same sample. Decreased water activity retards the growth of microorganisms, slows enzyme catalyzed reactions and, lastly, retards non-enzymatic browning. In contrast, the rate of lipid autoxidation increases in dried food systems.

Table 2.4. Water activity of some food

Food	a_w
Leberwurst	0.96
Salami	0.82-0.85
Dried fruits	0.72-0.80
Marmalades	0.82-0.94
Honey	0.75

Foods with a_w values between 0.6 and 0.9 are known as "*intermediate moisture foods*" (IMF). These foods are largely protected against microbial spoilage.

One of the options for decreasing water activity and thus improving the shelf life of food is to use additives with high water binding

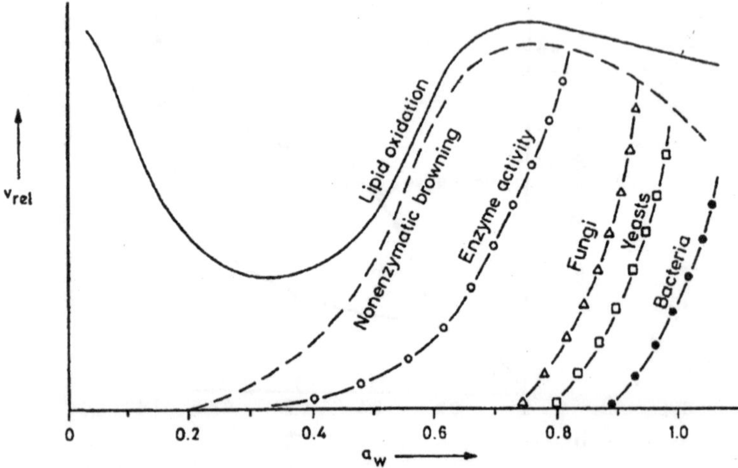

Fig. 2.4. Food shelf life (storage stability) as a function of water activity.

capacities (humectants). Table 2.5 shows that in addition to common salt, glycerol, sorbitol and sucrose have potential as humectants. However, they are also sweeteners and would be objectionable from a consumer standpoint in many foods in the concentrations required to regulate water activity.

Table 2.5. Moisture content of some food or food ingredients at a water activity of 0.8

	Moisture content (%)
Peas	16
Casein	19
Starch (potato)	20
Glycerol	108
Sorbitol	67
Saccharose	56
Sodium chloride	332

Water Activity as an Indicator

Water activity is only of limited use as an indicator for the storage life of foods with a low water content, since water activity indicates a state that applies only to ideal, i.e. very dilute solutions that are at a thermodynamic equilibrium. However, foods with a low water content are non-ideal systems whose metastable (fresh) state should be preserved for as long as possible. During storage, such foods do not change thermodynamically, but according to kinetic principles. A new concept based on phase transition, which takes into account the change in physical properties of foods during contact between water and hydrophilic ingredients, is better suited to the prediction of storage life.

Phase Transition of Foods Containing Water

The physical state of metastable foods depends on their composition, on temperature and on storage time. For example, depending on the temperature, the phases could be glassy, rubbery or highly viscous. The kinetics of phase transitions can be measured by means of differential scanning calorimetry (DSC), producing a thermogram that shows temperature T_g as the characteristic value for the transition from glassy to rubbery (plastic). Foods become plastic when their hydrophilic components are hydrated. Thus the water content affects the temperature T_g, for example in the case of gelatinized starch.

During the cooling of an aqueous solution below the freezing point, part of the water crystallizes, causing the dissolved substance to become

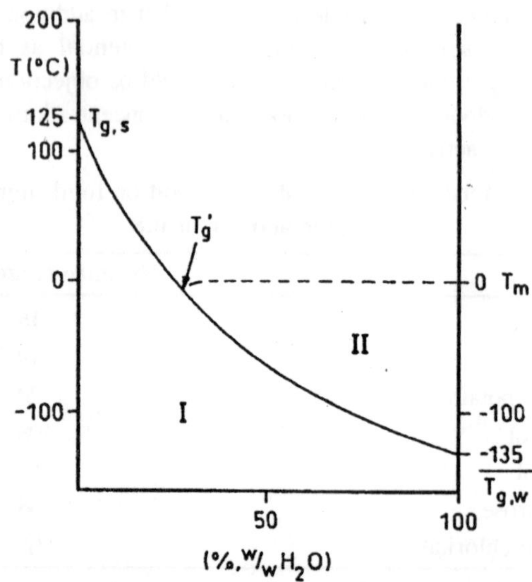

Fig. 2.5. State diagram, showing the approximate T_g temperatures as a function of mass fraction, for a gelatinized starch-water system.

enriched in the remaining fluid phase (unfrozen water). In the thermogram, temperature T'_g appears, at which the glassy phase of the concentrated solution turns into a rubber-like state. The position of T'_g (–5 °C) on the T_g curve is shown by the example of gelatinized

Table 2.6. Phase transition temperature T_g and melting point T_m of mono- and oligosaccharides

Compound	T_g *[°C]*	T_m
Glycerol	-93	18
Xylose	9.5	153
Ribose	-10	87
Xylitol	-18.5	94
Glucose	31	158
Fructose	100	124
Galactose	110	170
Mannose	30	139.5
Sorbitol	-2	111
Sucrose	52	192
Maltose	43	129
Maltotriose	76	133.5

starch; the quantity of unfrozen water W'_g at this temperature is 27% by weight. Table 2.7 lists the temperatures T'_g for aqueous solutions (20% by weight) of carbohydrates and proteins. In the case of oligosaccharides composed of three glucose molecules, maltotriose has the lowest T'_g value in comparison with panose and isomaltotriose. The reason is probably that in aqueous solution, the effective chain length of linear oligosaccharides is greater than that of branched compounds of the same molecular weight.

Table 2.7. T'_g and W'_g of aqueous solutions (20% by weight) of carbohydrates and proteins

Substance	T'_g	W'_g
Glycerol	-65	0.85
Xylose	-48	0.45
Ribose	-47	0.49
Ribitol	-47	0.82
Glucose	-43	0.41
Fructose	-42	0.96
Galactose	-41.5	0.77
Sorbitol	-43.5	0.23
Sucrose	-32	0.56
Lactose	-28	0.69
Trehalose	-29.5	0.20
Raffinose	-26.5	0.70
Maltotriose	-23.5	0.45
Panose	-28	0.59
Isomaltotriose	-30.5	0.50
Potato starch (DE 10)	-8	
Potato starch (DE 2)	-5	
Hydroxyethylcellulose	-6.5	
Tapioca (DE 5)	-6	
Waxy corn (DE 0.5)	-4	
Gelatin	-13.5	0.46
Collagen, soluble	-15	0.71
Bovine serum albumin	-13	0.44
α-Casein	-12.5	0.61
Sodium caseinate	-10	0.64
Gluten	-5 to -10	0.07 to 0.41

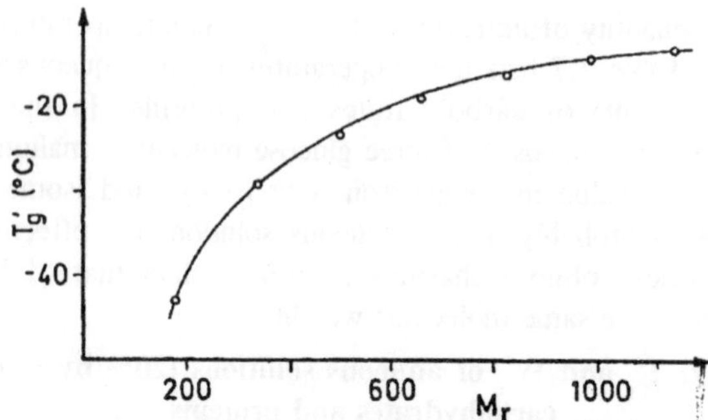

Fig. 2.6. Phage transition temperatures T'_g (aqueous solution, 20% by weight) of the homologous series glucose to maltoheptaose as a function of molecular weight M_r.

In the case of homologous series of oligo- and polysaccharides, T_g and T'_g increase with the molecular weight up to a certain limit. Table 2.8 lists the phase transition temperatures T'_g of some fruits and vegetables.

Table 2.8. Phase transition temperature T'_g of some fruits and vegetables

Fruit/vegetable	*T'_g(°C)*
Strawberries	–33 to -41
Peaches	-36.5
Bananas	-35
Apples	-42
Tomatoes	-41.5
Peas (blanched, frozen)	-25
Carrots	-25.5
Broccoli, stalks	-26.5
Broccoli, flower buds	–11.5
Spinach (blanched, frozen)	–17
Potatoes	–11

WLF Equation

The viscosity of a food is extremely high at temperature T_g or T'_g (about 10^{13} Pa.s). As the temperature rises, the viscosity decreases, which means that processes leading to a drop in quality will accelerate. In the temperature range of T_g to about (T_g + 100 °C), the change in viscosity does not follow the equation of *Arrhenius*, but a relationship formulated by *Williams*, *Landel* and *Ferry* (the WLF equation):

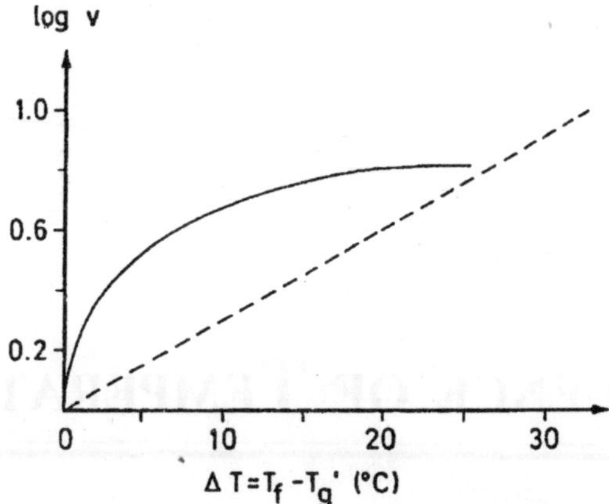

Fig. 2.7. Crystallization of water in ice cream.

$$\log \frac{\eta}{\rho T} / \frac{\eta_g}{\rho_t T_g} = -\frac{C_1(T - T_g)}{C_2 + (T - T_g)}$$

Viscosity (η) and density (ρ) at temperature T; viscosity (η_g) and density (ρ_g) at phase transition temperature T_g; C_1 and C_2: constants.

According to the WLF equation, the rate of which in our example water crystallizes in ice cream at temperatures slightly above T'_g rises exponentially. If the *Arrhenius* equation were to be valid, crystallization would accelerate linearly at a considerably slower rate after exceeding T′g.

In summary, we find that the rate of a food's chemical and enzymatic reactions as well as that of its physical processes becomes almost zero when the food is stored at the phase transition temperature of T_g or T'_g. Measures to improve storage life by increasing T_g or T'_g can include the extraction of water through drying and/or an immobilization of water by means of freezing, or by adding polysaccharides.

3

INFLUENCE OF TEMPERATURE

As early as the beginning of the present century the adverse effects of high temperatures on scrotal mammals appear to have been suspected in cryptorchid men. Since that time there have been many observations on the effect of high temperatures on the reproductive capacity of a wide group of plants, insects, lower vertebrates, and scrotal mammals.

Cowles in his excellent review relating hyperthermia to aspermia, mutation rates, and evolution has pointed out that there is a narrow temperature range which seems to be suitable for male gametogenesis in a wide range of plants and animals. Nonlethal temperatures above this range tend to prevent gamete formation or heighten mutation rate. This range tend to prevent gamete formation or heighten mutation rate. This temperature effect seems to be peculiar to gametogenesis in the male; female plants and animals are not affected. Such nonlethal temperature effects in modern reptiles have prompted at least one group of researchers to suggest that heat-induced sterility of the male may have been involved in archosaurian extinction. Clearly, if mutation and sterility result from increasing temperatures before somatic functions are impaired, then these reproductive changes would be of greater evolutionary significance than death or damage to the individual during periods of gradual temperature change.

Plants and poikilothermic animals exhibit a number of temporal or behavioural characteristics that seem to be effective in protecting gametogenesis from damaging temperatures. In all these organisms it is the temperature of the immediate environment that determines the temperature of the body and reproductive processes. In many species the timing of gamete production has evolved to occur at a time during the day or the year when heat damage is most likely to be minimal. Where a choice of temperatures is available to an organism, as in deep water or in burrows and crevices, the temperature limitations of gametogenesis may have been an outstanding factor in the evolution of preferred temperatures.

The success of homoiothermy in the animal kingdom testifies to the selective advantage gained from temperature control and maintenance of the highest possible temperature in the internal environment. However, from the temperature considerations discussed above, it would seem that the increase in somatic effectiveness could only be gained at the cost of reproductive impairment, unless gametogenesis in the male came under the protection of some temperature regulating mechanisms that were independent of the rest of the organism. Just such a testis-specific temperature regulation is seen in the scrotum present in almost all mammalian species. In birds, such a mechanism, if present, is less distinct. Although a few of the avian species have shown evidence of thermal protection for spermatogenesis, it is possible that many birds and some of the nonscrotal mammals have overcome or circumvented the temperature sensitivity of male gametogenesis.

In the scrotal mammals many conditions have been found to affect the spermatogenic function. Season and humidity, as they affect environmental temperature and heat regulation, are well-established factors. Hyperpyrexia due to systemic illness or local inflammation in the testis and scrotum produce degenerative effects on the spermatogenic function. Increased temperatures resulting from hot baths or scrotal insulation from clothing in man or from wool covering the scrotum in sheep produce the same effect. The effect of cryptorchidism (the retention of the testes in the abdominal cavity) is also well known. In addition to these circumstantial causes of temperature fluctuations that particularly affect the scrotal testis, there is a number of experimental techniques which change the temperature of that organ and have been shown to prevent spermatogenesis.

The effects of heat on the male were first observed as an interruption of fertility. Closer examination revealed gross anatomical changes as shown by changes in the size, weight, and texture of the testes. Further examination showed that cytological changes had occurred. Still more recent studies have demonstrated a number of physiological and metabolic effects including changes in blood and lymph flow, gas tensions, testicular fluids, specific metabolic pathways, and enzyme systems. These changes, not doubt, affect the endocrinological state as indicated by secondary sex characteristics and the usual expected role of testosterone in maintaining the reproductive tract.

In contrast to the effects of heat on testicular function, extensive short-term reductions in temperature can be tolerated without serious effects on the production of sperm and fertility in a wide range of species. However, continuous cold and the other changes usually concurrent with cold will stop spermatogenesis in some species.

The Effect of Heat

Fertility, Semen Quality, and Spermatogenesis

Some of the more specific effects of heat on fertility, semen quality, and spermatogenesis are considered here because these seem to be the most generally affected and are the end result of the application of heat, whatever the source. Cowles (1965) has reviewed the effects of heat on insects as it produces mutagenic effects and even sterility. In vertebrates other than mammals the experimental application of heat has usually involved the whole body. As all, except birds, are cold-blooded, the environmental temperature is a factor of prime importance in reproduction of these animals. In mammals whole-body and local heating of the scrotum have been studied. The end results of both appear to be similar.

Vertebrates Other than Mammals

In fish, amphibians, and reptiles temperature is often the signal for the timing of the breeding season and the beginning of spermatogenesis. Temperature frequently affects the rate of spermatogenesis, modified circadian rhythms, and may even produce heat sterility. Although light is the principal environmental factor affecting reproduction in birds, photoeffects are often modified by temperature in bringing about seasonal breeding. In addition, diurnal fluctuations in body temperature may prove to be an important mechanism for spermatogenesis in the homoiothermic avian species.

Fish

Photoperiod and temperature are the prime environmental regulators of the reproductive process for teleost fishes of the temperate and frigid areas. However, the relative importance of the two factors varies markedly in different species. For example, in the brook trout, *Salvelinus fontenalis*, and the Japanese killifish, *Oryzias latipes*, the photoperiod is the most important factor, whereas in the male minnow, *Covesius plumbeus*, temperature is the major factor. The effects of these environmental regulators vary with the stage of gonadal maturation, and the gametogenic processes appear to be independent of these environmental factors at certain stages of maturity. In addition, the effect may be different in males and females of the same species. In the tropics temperature may be of little importance as a regulator, rainfall and flooding being the primary regulating factors.

In some fishes sperm production takes place in response to rising temperature with little or no spermatogenesis occurring in cold weather. This has been shown by Geiser (1922) in the top minnow, *Gambusia affinis*, by Burger (1939) in the killifish, *Fundulus heteroclitus*, and by Merriman and Schedl (1941) in the American four-spined stickleback, *Apeltes quadracus*. Other fishes that respond to increasing temperature with reduced gonadal maturation, the "spring breeders," include the European bitterling carp, pike, darters, and sticklebacks.

A number of species respond to temperature higher than 15^{0}C with an increased rate of spermatogenesis, although spermatogenesis only occurs in the winter. For example, in the three-spined stickleback, *Gasterosteus aculeatus*, in seasons when the decline in water temperature as slow, spermatogenesis was completed early, and the fish reached potential maturity. Potential maturity changed to functional maturity in these individuals as a result of the rise in temperature in the spring. In the lakechub, *Covesius plumbeus* there is some evidence for an endogenous rhythm synchronized by temperature. Variations in light were not effective in reducing the stimulating effect of low temperature on spermatocyte formation, although short photoperiods did enhance it. However, complete maturation of the testes could be brought about experimentally only in individuals treated with daylight lengths comparable to natural conditions. Temperatures of 5^0-12^{0}C were essential for normal gonadal proliferation and primary spermatocyte formation and temperatures of 16^0-21^{0}C were necessary to promote proliferation of spermatogonia and bring on the completion of sperm formation in prespawning fish. Decreasing temperature induced gonadal maturation in the char, lake trout, and Caspian sturgeon. These are frequently referred to as autumn breeders.

In some fishes both water temperature and light control the gonadal cycle in the male. For example, during winter a higher temperature alone stimulated early spermatogenesis in the minnow, *Phoxinus laevis*, whereas light had no effect unless temperature was about 17^{0}C. In the viviparous sea perch, *Cymatogaster aggregata*, early winter daylength did not produce spermatozoa when temperature was kept at 10^{0}C. At 20^{0}C sperm were produced regardless of daylength. In late winter and spring which is the normal season of reproductive activity for this species, daylength became more important so that in short photoperiods at 20°C only 15% of the cells were beyond primary spermatocytes, whereas in a long photoperiod 46% of the cells developed to the spermatozoan stage. In late spring short photoperiods caused an involution of the testis.

More specific delineation of the effects of temperature have been shown in the Japanese killifish, *Oryzias latipes*. Autoradiography with thymidine-^{3}H showed that increased temperatures from 15°C increased the rate of development of all stages in spermatogenesis from the leptotene spermatocytes. At 15°C 12 days were required to complete the stages from spermatocyte to early spermatid, whereas at 25°C only 5 days were required. To go from the spermatid to spermatozoan stage at 15°C, and 8 days were required, whereas at 25°C only 7 days were required. Similar stimulation to spermatogenic function was shown in the guppy, *Poecilia reticulata*, where strict thermal limits were shown for maximum germ cell generation. Below 15°C and above 30°C degeneration occurred.

Amphibians

The spermatogenic cycle in some amphibians has been reviewed by Van Oordt (1960). In some Urodella, temperature seems to be the principal factor influencing the spermatogenic cycle with temperature below 12°C causing the cessation of spermatogenesis and degeneration of the spermatocytes and spermatids, whatever the season. Winter temperatures of 12°-17°C induced spermatocyte formation, and higher temperatures of 21°-25°C in winter caused a discharge of sperm and initiation of a new cycle. Neither light nor starvation affected this cycle. In the four-toed salamander spermatogenesis was made to continue throughout the whole year by maintaining a constant temperature.

Spermatogenesis in the toad, *Bufo bufo*, continued longer when temperatures were above 15°C. In the tropical toad, *Bufo melanostictus*, spermatogenesis was continuous despite seasonal variations of temperatures from 38° to 15°C and experimental cold of 10° to 15°C for 30 days. In the European frog, *Rana temporaria*, spermatogenesis was discontinued with high temperatures of 20°-24°C for 4-8 weeks, because the secretion of LH was impaired, and interstitial cells regressed, while androgen secretion was reduced and pituitary cytological changes occurred. However, similar temperatures tended to stimulate FSH production and new spermatogenic activity was initiated. Similar relationships between temperature and the gonad were found in the edible frog, *R. esculenta*. Van Oordt and Lofts (1963) explained this internal rhythm which regulates the sensitivity of the germinal epithelium to gonadotropin and the seasonal refractoriness of the primary spermatogonia to this hormone as the factor which prohibits the transformation of the discontinuous spermatogenic cycle of the frog into a continuous cycle.

Reptiles

The effects of photoperiod and temperature on testicular function in several species of lizard has been reviewed by Licht (1967a). Despite the existence of a distinct photoperiodism, daylength alone does not account for the seasonal testicular cycles in several species of lizard. Only under certain thermal conditions can photoperiodism initiate testicular activity, and the level of body temperature seems then to largely responsible for control.

In the American chameleon, *Anolis carolinensis*, temperature seemed to be more important than daylength for the rate of testicular growth. Thus, in this species no testicular development occurred at temperatures below 20°C; from 20°C upward growth, but retarded spermatogenic development, took place. The preferred temperature of 32°C was optimal for spermatogenesis and androgen production. Although higher temperatures accelerated testicular growth and development in this species after spermatogenesis was initiated, these same temperatures before initiation of spermatogenesis caused a delay in the onset of activity. Thus, the testes were refractory to high autumn temperatures until after mid-October when these temperatures were less likely to occur or exposure time was reduced. Daylength apparently acted only as a controlling mechanism in permitting temperature effects, which may be a mechanism preventing premature spermatogenesis in this species.

At 35°C the American chameleon exhibits testicular damage. In contrast to this, the desert sand lizard, *Uma scoparia*, preferred a temperature of 37°C, and testicular damage occurred at 40°C. While the preferred temperature of the lizard may be suitable for testicular function, nonetheless exposure to slightly higher temperatures caused a loss of appetite and body weight. Therefore, it is unlikely that the preferred temperature of these animals is governed exclusively by requirements for optimal testicular function. A further complicating factor is that testicular degeneration may occur even at preferred temperatures if these temperatures are prolonged. McGinnis (1965) has shown that western fence lizards, *Sceloperus occidentalis*, held at their preferred body temperature for either 12 hr/day or 24 hr/day over 2 weeks suffered partial or nearly complete degeneration of testis (based on weight and size). As controls maintained the large testes characteristic of the spring condition, it is apparent that frequent intervals at temperatures lower than "preferred" are necessary to the normal function of the testis at least in this species.

Some species of lizard, like *Phrynosoma cornutum,* have long periods of testicular quiescence, and during such quiescent periods responded little, whether kept in the dark at 8⁰C or in the light at 35⁰C during mid-winter. *Sceloporus occidentalis* showed greatest testicular weight in the coolest months and lowest testicular activity during hot months. Cowles (1965) indicated that this situation represents the general condition for the reproductive season in most lizards of the southwestern United States. The American chameleon was an exception to this, as spermatogenesis extended throughout the summer.

In snakes the essential part of spermatogenic activity takes place after hibernation. In the adder, *Vipera berus*, spermatogenesis began following the cessation of hibernation in March and April and varied according to climatic conditions. Temperature especially affected the rate of spermatogenesis with cooling slowing down the rate and heat almost in stantly renewing the activity in the germinal epithelium. A similar situation is reported in the garter snake, *Thamnophis elegans terrestris*, where in the northern hemisphere the first mature sperm were released in the tubules in early June, and the testes were found to reach their maximum activity in early July and August with a decline coming in September and October.

Birds

In the white-crowned sparrow, *Zonotrichia leucophrys* nuttali, Blanchard (1941) found a variation in the date of the onset of spermatogenesis to be due to the number of days above 5⁰C. Although other climatic factors may be involved, temperature was found to be the principal factor. In another variety of white-crowned sparrow, Z. *leucophrys* gambelii, high environmental temperature was found to augment the gonad-stimulating effect of artificially long days. However, in this species daylength was the more potent single factor involved in testicular development. In the starling, when spermatogenesis was light-induced, Burger (1948) found that testis volume was larger at 32.2⁰C or 37.8⁰C than at 11.1⁰ to 22.2⁰C. Under a light regime which did not induce spermatogenesis, a constant 32.2⁰C failed to produce progressive testicular enlargement in this species. In several species of passerine birds, Marshall (1949) reported that the severity of winter played an important part in the timing of the breeding season.

The rooster of the domestic chicken seems to be the exception to the general rule of sperm production in most avian species, for sperm are produced throughout the year. In the northern hemisphere increased spermatogenesis was found to be associated with increasing light between December and April. In this species testes developed normally throughout a considerable range of lower temperature.

Exposing White Plymouth Rock cockerels during a 3-hr period to heat stress (38^0-40^{0}C), which caused death to several, consistently lowered sperm concentration and semen volume, but these values returned to normal 5-6 days later. On the other hand, in White Leghorns exposure to 9 hr of 38^{0}C followed by 15 hr at 21^{0}C had no effect on volume, concentration, or metabolism of the sperm produced, except in a strain that had been developed in a cool environment. In the turkey, the testes of yearling birds shrink during the winter months indicating that spermatogenesis is not occurring. Fertility in turkeys was improved in winter by housing them in warm pens at 18.3^{0}C, when minimal outside temperatures were -29^{0}C. Cooling to 18.3^{0}C also improved fertility in summer, when maximum outside temperatures at times reached 37.7^{0}C. In this species also prolonged elevation of body temperature resulted in damage to the germinal epithelium of the testis.

These observations on domestic birds suggest that heat sensitivity of spermatogenesis in the bird may be similar to that observed in mammals and some cold-blooded species. The possibility that the peculiar heat sensitivity of male gametogenesis is also present in birds is further supported by the work of Williams (1958). In these experiments on the rooster, testes transferred to a lateral subcutaneous position (where the temperature was 39.0^{0}C), developed faster than in the normal deep body position (41.3^{0}C).

Riley (1937) described how some of the earlier studies with the house sparrow, *Passerine domesticus*, failed to demonstrate mitotic activity in the testes. Riley showed a diurnal spermatogenic cycle during the breeding season in this species with activity at a peak between 2 and 4 a.m. when body temperature was down to 40^{0}C from 43^{0}C. During the day, few if any mitotic figures could be found. Forced exercise during the night with consequent higher temperatures also retarded spermatogenesis. It is possible that this diurnal periodicity in the spermatogenesis of birds is an important mechanism of protection from thermal damage. If the heat-sensitive stage of spermatogenesis is confined to a certain stage of germ cells formation as many mammalian studies indicate, then limitation of this critical stage to a time when body temperatures are lowest could overcome the problem in birds. It is important that these studies be extended to other species of birds, so that the question of thermal sensitivity of avian spermatogenesis can be resolved.

Mammals

In all animals up to and including reptiles and birds, the testes lie high in the abdominal activity. Even in some of the more primitive mammals, such as Monotremata (duck-billed platypus and spiny anteater), this condition exists. Most of the Marsupialia carry their testes in a scrotum. Most of Edentata (sloth, anteater, and armadillo), some of Insectivora (shrew), Cetacea (whale and porpoise), Sirenia (seacows), Proboscidea (elephant), and Hyracoidea (hyrax) have testes lying in the abdominal cavity also. In some Rodentia, some Insectivora (mole and hedgehog), and Chiroptera (bats) the testes are found in the abdominal cavity during the resting phase of the sexual cycle, whereas in the mating season the testes descend into the scrotum. In Ungulata, Carnivora, and Primates the testes are extraabdominal and after puberty remain in the scrotum at all times.

In the species with scrotal testes, testicular function may be affected not only by factors which affect whole-body heat, but also by heat applied directly to the scrotum, insulation of the scrotum or even confinement of the testis to the body cavity as occurs in cryptorchidism. Changes in body temperature were not as effective in raising the temperature of the testis as was the application of local heat. On the other hand, whole-body heat is more likely to affect other body functions (endocrine, respiratory, circulatory), which may in turn affect the testes. The direct application of heat to the testis may also affect the gonad-hypothalamus-pituitary axis.

In mammals the effect of heat on testicular function seems to be similar regardless of the source or means of the application. High seasonal temperatures, which may be enhanced by high humidity, the direct application of heat to the testis by insulating the testes or translocating it into the body cavity tends to produce what Fukui (1923) called the "heat testicle." This term has since been dropped.

In different species the effect of light and temperature varies considerably, and in many cases an interaction may be responsible for the breeding cycle. In the golden hamster, *Mesocricetus auratus*, low temperature (6⁰C) and a 2-hr photoperiod brought about weight loss and testicular atrophy, and a temperature of 20⁰C with a long photoperiod induced a continuous breeding condition. In the European hamster, *Cricetus cricetus*, a temperature of 20⁰C was not effective in preventing the regression of gonads at the end of the breeding season. Many other species are affected by light as well as heat. However, in the ground squirrel, *Citellus tridecimlineatus*, testicular regression is associated with rising summer temperatures, and an experimental treatment at 4⁰C applied during the breeding season permitted the maintenance of the breeding condition during the whole year. In this case, light had no effect. The remainder of this discussion also has been restricted to the direct effects of elevated temperature on several species.

Although the adverse effects of heat on scrotal mammals were suspected many years earlier, it remained for Crew to postulate the effect of elevated temperature in producing the aspermatic condition of the testicle located in the abdomen, and for Moore and Fukui to demonstrate the effects of directly applying heat to the testes of guinea pigs, rats, rabbits, cats, dogs, goats, and men. The early papers by these investigators present an excellent review of the background information form which they designed their original experiments. They also demonstrated that recovery from the degenerative state produced by heat was possible. Various methods of applying heat, including warm water, warm air, warm paraffin, sunlight, arclight, and producing artificial cryptorchidism, were tested by these early investigators. A more complete description of some of the earlier investigations as well as many since that time is considered in the following as it applies to particular species.

Guinea Pig

In his early work, Moore (1924a) demonstrated that confining the testis of the guinea pig in the abdomen led to dissolution of the spermatogenic cells within 2 weeks. Moore (1924b) also demonstrated that a marked degeneration of the testis occurred within a few days if water at 6⁰-7⁰C above body temperature was applied to the scrotum for only 10 min. Other forms of heat, including heat emitted from a hot plate, a light bulb, diathermy currents, the application of hot saline on the exposed testes all produced degeneration. Continuous heat was more effective in causing damage than intermittently applied heat at the same temperature.

Young (1927) found that an application of water at 46^0-47^{0}C to the scrotum of the guinea pig for 15-30 min caused degeneration which was complete in 12 days. During these investigations it become apparent that the regenerative capabilities rested in the spermatogonia. The regenerative period was approximately 45 days in some of the tubules, and the number of sterile matings resulting from treated animals reached a peak between the 45th and the 75th day.

Phillips and McKenzie (1934) reported that the degeneration of spermatozoa in the isolated epididymis occurred no more rapidly when the epididymis was placed in the body cavity than in the normal scrotal position. From this they concluded that the effect of heat on fertility was due to the effects on the spermatozoa during the process of spermatogenesis and not on the fully formed sperm stored in the epididymis.

Moore (1926) made testicular grafts subcutaneously, intramuscularly, and intraperitoneally, and found that although the grafts which recovered contained cells in division, none were capable of reaching the final stage of spermatozoan differentiation.

Rat

The rat seems able to adapt to prolonged periods of heat near body temperature. No infertility or testicular damage was demonstrated by Harrison and Harris (1956) in rats maintained for prolonged periods at near body temperature (36.7^{0}C); no damage was observed at temperatures slightly higher than body temperature, even though death frequently occurred if the exposure was continued for as much as 40 hr at 39^{0}C. Exposure of rats to 18^{0}C resulted in the production of testes larger than when rats were exposed to 35^{0}C, and these results indicated that the higher temperature not only produced partial atrophy but also resulted in the presence of a higher level of gonadotropins.

The direct application of heat from several sources the rat testes, including infrared radiation (48^{0}C), hot air, and experimental cryptorchidism, all produced testicular degeneration and infertility. Ten 2-min exposures to a water immersion of the testis at 44^0 ± 1^{0}C daily for 8 weeks reduced testis weight but was not as effective as 8 weeks of cryptorchidism.

Experimental cryptorchidism resulted in complete infertility, whereas heat treatment by water rendered 11 out of 12 males infertile (Venkatachalam and Ramanathan, 1962). Bowler (1967) found that hot water treatment at 43.5^0 ± 0.1^{0}C for 20 min caused degeneration that required 52-65 days recovery time after a single application. The mean recovery time required after 36 weeks of 6 weekly heat treatments was 108 days. As the duration of spermato-genesis in the rat is 52 days, one heat treatment reduces the germinal epithelium to stem cell spermatogonia, whereas repeated heating probably increases the number of permanently damaged tubules and thus increases the time necessary to attain larger numbers of sperm.

As in the guinea pig the length of the infertile period in the rat following cryptorchidism or heat treatment appears to be dependent upon the duration of exposure as well as the temperature of exposure. Turner (1938) demonstrated the production of spermatozoa by testicular grafts into the anterior chamber of the eye where the temperature was found to be 30.6°C in male or female, normal or castrated rats.

Mouse

The mouse appears to be more sensitive to elevated environmental temperatures than the rat. Hardberger (1950) found that an air temperature of 32.2°C for 10 days resulted in infertility in the mouse with degeneration of the germinal epithelium by the 15th day. A temperature of 30°C prevented normal weight gain in the testis and seminal vesicles and prevented the normal development of the germinal epithelium in immature mice in studies conducted by Maqsood and Reineke (1950). Under these conditions, supplying thyroprotein improved spermatogenesis in the mouse.

Okauchi (1963) subjected mice to a temperature of 33°C for period of 5, 10, 15, and 20 days, respectively. Following heat treatment the animals were kept under conditions where the temperature ranged from 8° to 18°C for 60 days. At the end of each heat treatment and following the subsequent 60-day recovery period the percentages of abnormal spermatozoa, semen volume, and infertility were assessed. The proportion of abnormal sperm following 5 days treatment was approximately 10% as found in the control group. After 10, 15, and 20 days the percentages of abnormal spermatozoa increased to 33, 45, and 87%, respectively. Semen volume decreased also proportionately to the length of heat exposure. Following the 60-day recovery period the percentage of abnormal spermatozoa from mice previously exposed to the high temperature for 10, 15, and 20 days had decreased to 18, 26, and 39%, respectively. The major abnormalities consisted of sperm broken at the junction of the head and neck and various deformed head, midpiece, and tail structures. Following a 100-day recovery period, all males except two exposed to 20 days of the high temperature showed normal fertility.

Rabbit

The rabbit has been used extensively to study the effects of heat on the testes. Continuous high temperatures (32.2⁰C-33.3⁰C), even when alternated with lower temperatures (12.8⁰C-18.3⁰C) every 12 hr, have been shown to have a detrimental effect on the reproductive capacity of the male rabbit. These conditions produced two semen volume, poor sperm motility, a decreased concentration of spermatozoa and a reduced percentage of live and normal sperm. The effect of continuous heat was more pronounced than the effect of intermittent high and lower temperatures. Under the continuous heat regime, thyroprotein brought about further detrimental effects on the semen characteristics, but improved the semen characteristics under intermittent heat treatment. That the rabbit also can withstand fairly high temperatures for a short period of time without an effect on sperm motility or concentration, was demonstrated with a 30-120 min exposure of 43⁰C (relative humidity 30-40%) as shown in studies by El Sheikh and Casida (1955). Higher temperatures for 30 min did cause a marked decrease in spermatozoa in two animals that survived the treatment.

Some of Fukui's early studies (1923) were conducted with the rabbit. He demonstrated that 100 hr of retention of the testis in the abdomen produced a marked regression. He applied cooling from outside the body to a testis placed in the inguinal canal and demonstrated that degeneration did not occur as occurred in the opposite testis placed in the inguinal canal without cooling. He performed similar experiments on the dog. Asdell and Salisbury (1941) demonstrated that the extreme fertile life of spermatozoan in the rabbit which has had the testes placed in the abdominal cavity was 8 days. Motility of the sperm ceased at about 14 days. Their investigations also demonstrated that about 1 week was required for recovery of each layer of developing sperm cells in the spermatogenic process in the testis. Knaus (1958) demonstrated that the fertilizing capacity of sperm from the caudal epididymis following artificial cryptorchidism was lost after 4 days. However, sperm motility was still found after 8 days, but was completely absent after 12 days in the abdomen. The changes in seminal characteristic which follow experimental cryptorchidism are similar to those found with scrotal insulation in larger animals in which coiled tails appeared from 3 to 5 days following the translocation into the body cavity and decapitated sperm appeared after 9-11 days. Cryptorchidism produced dead and decapitated sperm in the isolated tail of the epididymis after 9 days with 4% more dead cells than in the control epididymis. In a more precise study of the effect of cryptorchidism on the caudal epididymal sperm, Glover (1960) found that abnormalities of the spermatozoa began to appear at 7 days. The effects were present throughout the whole length of the caudal epididymis and also in the vas deferens. Sperm in the caput epididymis were more susceptible than those found in the body or caudal epididymis. Rabbit spermatozoa incubated at 40^{0}C were still able to fertilize ova, but the resulting embryos were less likely to survive.

Cummins (1968) studied the effect of mating with cryptorchid males on implantation sites and the presence of normal fetuses 29 days later. Implantation sites as a percentage of the corpora lutea (CL) present for animals mated to males from 0 to 7 days following translocation of the testis to the abdominal cavity decreased from 73.6% to 0. Normal fetuses present expressed as a percentage of the CL were 30, 63, 33, 5, 28, 0, and 0, respectively. These data demonstrate that the failure of the ability to fertilize was first an effect of cryptorchidism on spermatozoa. This did not appear to be related to motility or the incidence of dead or abnormal spermatozoa, but was associated with an increased incidence of retained, expanded protoplasmic droplets, and with a reduction in the proportion of live, morphologically normal types. The suggested evidence of an improved fetal survival on day 1 as compared with the controls in the Cummins (1968) data has been suggested by the author as due to an accelerated aging, thus eliminating less "effective" spermatozoa. Salisbury ad Flerchinger (1967) have reported a similar improved fertility with ejaculated bull spermatozoa aged in storage at 5^{0}C for 1 day, or stored in frozen state for 4 months.

Dog

One of the early realizations that the testes would lose their spermatogenic activity if replaced in the abdominal cavity was reported by Griffiths (1893) using the dog. Fukui (1923) also used the dog in his early work on the effects of heat on the testis. Scrotal sac temperatures of 38^0-40^{0}C were produced with electric light bulbs by Chomiak *et. al*. (1954) in the dog to bring about reduced sperm activity in the ejaculate after 5 days. After 10 days exposure no activity was detected in the ejaculate. Motile sperm reappeared in 10 days after the original 10-day exposure and numerous motile sperm were present 15 days after cessation of the treatment. Thirty days were required for the reappearance of what appeared to be complete, normal productivity. This would appear to be due to an acclimatization to the treatment, for spermatogenesis was only temporarily inhibited by the techniques employed. Continuation of the heating treatment following the 5-day recovery period brought on diminished sperm numbers again, but after 20 days of treatment the sperm numbers were increasing. Cowles (1965) has suggested that the treatments used by these investigators may have permitted an accommodation by the scrotum allowing compensation of temperature within the narrow scrotal-abdominal gradient of the dog.

Cattle

A number of investigators have reported a decline in breeding efficiency in cattle during the hot months of the year. Concurrent with the low fertility, there have been reports that semen volume and sperm motility reached their lowest levels during the hot summer months with abnormal sperm 25% higher than during the spring months. Others have also reported a decline in semen quality of bulls in areas where there are hot summer months. A decline in the percentage of progressive motile sperm and the percentage of abnormal sperm resulting from summer temperatures has also been reported in spite of shade or air conditioning. Okamoto *et.al.* (1959) found that local cooling prevented seminal degeneration due to short periods of hot atmosphere. Milicevic *et.al.* (1968) found a significant effect on ambient temperature on the pH of semen, volume of the ejaculate, and the coefficient of survival time *in vitro,* but these effects were not serious at temperatures from 2⁰ to 21.6⁰C. These authors suggested the diurnal rhythm of temperature diminished the unfavorable influence of elevated ambient temperatures. In Mexico bulls reared in a hot room with the maximum temperature of 36⁰C for 8 hr from 26 weeks of age to puberty had the same degree of libido, although they showed impaired spermatogenesis, poor quality of semen, and low concentration of spermatozoa, compared with bulls reared at a temperature of 27⁰C maximum. Exposure of bulls to 40⁰C for 12 hr had a detrimental effect of spermatogenesis which was more severe in *Bos taurus* than in *Bos indicus*. Decreased semen quality was quality was observed with the maximum damage occurring 4 weeks after the exposure.

Casady *et. al.* (1953, 1955, 1956) kept bulls in environmental chambers where conditions of gradually increasing temperatures from a lower level of 22⁰ up to 37⁰C were compared to decreasing temperatures in another chamber from 30⁰ down to 11⁰C. Two weeks at 37⁰C or 5 weeks at 30⁰C impaired spermatogenesis, but sex drive or ejaculate volume were not affected. It was concluded by these investigators that spermatogenesis in the young dairy bull was impaired with continuous exposure to temperatures exceeding 29.4^{0}C for period exceeding 5 weeks.

Singleton (1968) found that innoculation of cattle with blood infested with *Babesia argentina* produced fever, and the affected animals had a resulting seminal degeneration and loss of fertility.

Sheep

Seasonal changes in semen production and the occurrence of summer sterility have long been known in sheep. McKenzie and Berliner (1937) found the lowest spermatogenic activity in rams during the summer months with spermatogenesis increasing starting in August to October and continuing through January. These results were found with the Shropshire and Hampshire breeds. With the Merino and British breeds of sheep in Australia seasonal changes in semen and fertility were observed which corresponded to atmospheric temperatures, with temperatures of 32.2^0-37.8^{0}C causing seminal degeneration. Moule (1948) checked the fertility of sheep in Queensland, Australia, and found that breeding effectiveness of range sheep was lowest in the hottest area. Bogart and Mayer (1946) also found summer sterility in sheep in the United States.

High atmospheric temperatures have been shown to lead to increased testicular temperature in the ram. Using scrotal insulation Phillips and McKenzie (1934) found that testicular temperature was raised to 37.0^{0}C compared to 34.9^{0}C for the uninsulated testis at a room temperature between 14^0 and 19^{0}C. Removal of the insulation resulted in a return to normal temperature within 20-30 min. Scrotal insulation brought about a decreased testis volume and as increase in the percentage of abnormal sperm, first observed in 1 week after insulation; 20 days after the application of the insulation the sperm supply was exhausted. A $2^1/_2$ months period of scrotal insulation left all seminiferous tubules in a state of disintegration in studies carried out by Moore and Oslund (1924). Glover (1955) found that a 24-hr scrotal insulation which increased testicular temperature by 5^{0}C resulted in the sudden appearance of semen with tailless spermatozoa 17-24 days after the treatment. Frequency of ejaculation had little effect on the time of appearance of abnormalities in the semen following scrotal insulation in rams.

The effect of heat, either environmental or applied, on semen quality in the ram has been investigated in a number of studies. Cupps *et. al.* (1960) found that the semen quality in rams in a hot central valley of California during the months of July through September showed decreased ejaculate volume, decreased sperm concentration, lowered sperm motility, and increased percentage of abnormal sperm. Fructose concentration of the semen, however, was not altered and there were no adverse effects on libido. Southdown rams subjected to 38°C for 24 hr showed declined sperm motility from 75 down to 30% motile sperm 1 week after treatment. Simpson (1966) also found similar semen quality deterioration with application of 38°C for 24 hr. Rathore and Yeates (1967) subjected Merino rams to two daily exposures of 8 hr at 40.5°C (45% relative humidity). This treatment produced abnormal sperm, beginning 8 days after the end of the treatment and reached a maximum of 19% abnormalities at 20 days following the treatment. Not all of the rams so treated were affected. Moule and Waites (1963) applied two 6-hr periods of 40.5°C over 3 days time to Merino rams and found a decrease in semen quality starting 13-21 days later.

The number of abnormal sperm increased, and the total fructose and fructose concentration of the semen increased. The maximum duration of the effect was 30-39 days. Cooling the scrotum with circulating water at 17°-19°C was sufficient to prevent degeneration. Continuous exposure of rams for 78 hr to a hot room at 38°C produced seminal degeneration in 15±3.7 days after the heating started. Lindsay (1969) used Merino, Dorset Horn, and Border Leicester rams to test the effects of continuous exposure to 27°-43°C. Only the Merino rams were able to maintain sexual activity at 43°C. The dorset Horn rams recovered when the temperature was lowered, but the Border Leicester rams did not. Semen quality and reaction time were both reduced by the higher temperatures.

Shearing and air conditioning (Dutt and Simpson, 1957; Whiteman and Brown, 1959) improved semen quality and fertility in rams. Foote *et.al.* (1957) found that when room temperature was increased from 21.1° to 35.0°C, increases in rectal (38.9°-40.5°C) and testicular (35.0°-37.0°C) temperature occurred in rams. However, shearing enabled the rams to maintain a testicular temperature 1°C below the unsheared rams. Application of high humidity (78-88%) resulted in a 1°C higher testis temperature than occurred with lower humidity (57-69%). Dutt and Hamm (1957) showed that a constant temperature of 33.5°C and 60-65% relative humidity for 1 week lowered sperm motility and general semen qualities in unsheared rams exposed to this temperature in the winter. Sheared rams were less damaged and treated rams returned to normal 8 weeks following termination of the heat treatment. Simpson and Rice (1957) found that a 32.2°C temperature caused a rectal temperature of 40.3⁰C and a testis temperature of 36.7⁰C, whereas control rams maintained a rectal temperature of 38.8⁰C and a testis temperature of 34.1⁰C. Dutt and Simpson (1957) applied air conditioning at 7^0 to 9^0C during summer months and significantly improved the semen quality and fertility level of Southdown rams. Breeding tests showed 64.2% fertilized ova present in ewes mated to the cooled rams as compared to 26% in ewes mated to control rams. The control group was kept in the shade so that they were exposed only to the ambient temperatures of 26^0-32^0C and not to solar radiation.

Domestic Pig

A marked drop in fertility during the summer in the boar has been attributed to the effect of summer temperatures. Summer conditions (up to 35^0C) reduced fertility significantly in swine when compared with animals kept under air conditioning. This was determined by the reduced farrowing rate of animals kept outside the air conditioned buildings. However, early embryonic death, as indicated by the number of animals returning to estrus at more than 45 days following the insemination (the normal estrous cycle in the sow is 21 days), were identical for sows inseminated with semen from boars from either group.

In studies of boars kept at a constant elevated ambient temperature (35^0C for 4-8 weeks) and local acute high temperatures (40.5^0C for 3 hr), Mazzarri *et. al.* (1968) found that semen volume was not affected in either case. However, the number of sperm, the percentage of motile sperm and fertility were affected much more drastically by the local applied higher temperature. Fertility reached 0 to 37-51 days following the treatment when both the number and percentage of motile sperm were also at a minimum. Mazzarri *et. al* (1970) have also shown that high air temperature (35^0C) plus a long daylength of 16 hr were more effective in lowering fertility of boars than was the effect of heat alone.

Camel

Volcani (1953) has reported summer sterility in the camel in Israel. The male camel is a seasonal breeder, and spermatogenesis is initiated during the period from November to January with the maximum activity form February to April and a disintegration of the cell layers and reduction in the size of the testis from May to July. During the period August to October the testes are 30% smaller, and the male shows no interest in the female. Although Volcani indicates that these effects are due to the prevailing temperatures during the summer, careful studies are needed to be certain of the relative effects of both heat and light.

Monkey

Adult male monkeys were placed in a special cage which resulted in the testes being immersed in a water bath at 44^0 ± 1^{0}C for a 20-min period. This treatment was repeated 6 days a week for a period of 8 weeks. Compared to controls the treated animals were found to have testes one-third to one-half normal size and an atrophy of the germinal epithelium.

Man

Although the mode of action of fewer-producing diseases in causing testicular necrosis was not understood by the early observers, typhus, scarlatina, tetanus, peritonitis, and the fever associated with pneumonia, typhoid, and chicken pox have all been associated with testicular deterioration in man. More recent reports on these fever-producing conditions have been made by others in the light of modern understanding.

A 10-year study of human fertility in the seasonally hot and humid region of Galveston, Texas, showed that conceptions were less frequent in July and August and most frequent in January, with 77% more conceptions in January than in August. This same difference existed for both Negroes and whites and no detectable correlation was found with any factors other than heat and humidity. Conception rate curves were found to be inverse to temperature but bore no relationship to the humidity alone.

MacLeod and Hotchkiss (1941) used artifically induced hyperthermia to raise body temperature to 41^{0}C in healthy young male donors. After 30 min in the fever cabinet (43^{0}C) body temperature was elevated 0.5^0-4^{0}C for up to 3 hr. The effects of the heat treatment on the testes did not become evident for nearly 3 weeks and reached a maximum, as indicated by the drop in the number of sperm produced, at approximately 42 days. Recovery required an additional 25 days. No marked changes in motility or morphology of the spermatozoa were seen. Use of the sauna bath 8 times in 2 weeks with a mean exposure time of 15.3 min was found to be sufficient to elevate body temperature 0.93^{0}C. The temperature of the bath ranged from 77^0-90^{0}C. An average maximal reduction in total sperm number occurred in 30-39 days following the bathings. Motility changes paralleled those of total spermatozoa. Recovery in semen quality came in approximately 10-15 days following the treatment.

Ehrenberg *et. al.* (1957) found a scrotal temperature difference of $3.3^0 \pm 0.4^0C$ between clothed and disrobed subjects. This difference between 34.0^0 and 30.7^0C indicates the effect that clothing can have on testicular temperature. Rock and Robinson (1965) found that scrotal insulation produced decreased sperm concentration and decreased total sperm in men. Later these same investigators found that the use of a modified athletic supporter decreased the temperature gradient by 0.8^0C. This small change every day depressed mean total sperm count by the 3rd week with a minimum being reached by the 7th week. After the treatment was stopped, recovery was slow for 9 weeks and then appeared to accelerate so that sperm concentrations rose above the pretreatment levels. They present evidence that longer treatment depressed sperm count even further. A similar rebound effect was achieved using a 150 W light bulb head near the scrotum for 30 min per day for 14 consecutive days.

Submerging the scrotum in water at 43^0-47^0C for 30 min daily for 6 or 12 consecutive days produced the expected fluctuation in sperm count and suppressed spermatogenesis. Discontinuation of the treatment was followed by apparently normal spermatogenesis. In this study, 2-3 weeks after heat treatment (either 6 or 12 consecutive days) there was a temporary rise in sperm count to 150-310% of the initial levels. After 9 weeks another sharp rise to 170-560% of the pretreatment levels occurred and lasted for 1-5 weeks.

Varicocele has been shown to be associated with hypospermia. Although this condition has been thought to be due to an elevation of testis temperature, this has been disputed.

There are a number of everyday possibilities for heat injury to the testes in man. Several of those already mentioned include hot baths tight clothing, a vocation or situation in which man is required to labor under hot conditions, fever, and even posture. Thus, man, probably more than any other species, is likely to be subjected to conditions which are detrimental to sperm production and fertility.

Specific Cellular Sensitivity

From the early experiments, every little was learned about the relative sensitivity of different cell types to heat, except that the more advanced cells appeared to be the most sensitive and spermatogonia the least. The reasons for this were twofold. A classification of the stages of the cycle of the seminiferous epithelium and clear recognition of all the cells, particularly the different types of spermatogonia, awaited the studies of Leblond and Clermont (1952a, b), Roosen-Runge

(1952), Clermont and Leblond (1953), and others. Without these careful analyses, errors in recognition of the origin of degenerating cells were often made. But in addition, the severity of the heat treatment often precluded any hope of observing differential sensitivity. Temperatures 6^0-10^0C above body temperature are not physiological, and even if applied for only a few minutes may have a nonspecific lethal effect, as pointed out by Steinberger and Dixon (1959). Indeed, the phospholipid components of all mammalian cells are in danger of irreparable damage if subjected to temperatures in the range 40^0-45^0C for any length of time. The seminiferous tubules need to be examined soon after heating if the precise stage of spermatogenesis vulnerable to heat are to be recognized.

Spermatogonia

Spermatogonia of the guinea pig and mouse were the most resistant germinal cells to artificial cryptorchidism and extreme local heating, and the spermatogonia of the pig were unaffected by temperature elevation. Some rat spermatogonia survived abdominal temperature longer than any other germ cell. By contrast, threshold doses of heat (41^0C for 15 min) were reported selectively to destroy spermatogonia in rats, although Steinberger and Dixon (1959) suspected that degenerating resting primary spermatocytes may have been mistaken for degenerating spermatogonia. Spermatogonia of rats were undamaged by exposure to 43^0C for 15 min. However, when the testes of rams were heated to 40^0C for about $2^1/_2$ hr, the number of B-types spermatogonia in the prophase stage of mitosis increased within 12 hr, and a four- to fivefold accumulation occurred within 24-48 hr. Many of these cells do not complete mitosis and are apparently blocked between prophase and metaphase. Some cross sections of tubules contained up to 30 spermatogonia in division, while others contained none. This was considered to be evidence for synchronization of cell division among spermatogonia. A_1-type spermatogonia were not affected by the heat treatment. Skinner and Louw (1966) reported "an apparent tendency for the number of spermatogonia in mitotic division to increase" after bulls were exposed to high air temperatures.

Despite the different heat loads applied, there may be a species difference in the thermal sensitivity of spermatogonia of rodents and ruminants. Alternatively, if the damaging action of heat on sheep testis is partly due to local tissue hypoxia, aggravated by the absence of an increase in testicular blood flow, then the elevated blood flow in rat testes at 43^0C may bring enough oxygen to maintain normal mitosis.

Spermatogonia in prophase enough oxygen to maintain normal mitosis. Spermatogonia in prophase and intermediate spermatogonia were reported to be among the first cells to be killed by periods of testicular arterial occlusion in rats, which could be evidence that mitosis is particularly oxygen-dependent. However, Hilscher (1964) was unable to recognize a pattern of damage among spermatogonia after temporary ischemia.

The differential thermal sensitivity of sheep spermatogonia, with the A_1-type being the most resistant, follows the pattern of susceptibility of mouse spermatogonia to low doses of X-irradiation. Their survival may account for the rapid reappearance of spermatozoa in semen 53-60 days after the end of short periods of testicular heating in severely affected rams and rats, and after 10-12 weeks in man. The extra cells may result from the division of A_0-type spermatogonia, which were recently described in rat and bull. These do not normally enter into division but are thought to be available as reserve stem cells for recruitment to rebuild the spermatogonial population after damage. They are resistant to heat. Such spermatogonia may account for the hypertrophy of the remaining testis after unilateral castration and rete testis cannulation in rams.

Spermatocytes

Certain stages of the long meiotic prophase of primary spermatocytes are specifically thermally sensitive in all species so far examined. Primary spermatocytes were the first cells to die following heating of the guinea pig testis and this cell type was also affected in the rat. When rat testes were heated to 43°C, primary spermatocytes from late stage VII pachytene to dividing spermatocytes at stage XIV, were destroyed; and these were the cells primarily affected by cryptorchidism.

Glucose supplementation greatly increased the incorporation of labeled lysine into the protein of cells in these stages so that limitation of supply of glucose may be involved. The pachytene spermatocytes of the sheep and pig pass through a thermally sensitive period toward the end of stage 7 and beginning of stage 8, and few survived if heated to 40°C for $2^1/_2$ hr in sheep or 40.5°C for 3 hr in the pig.

The pycnotic nuclei and nuclear debris, abundantly present 12-24 hr after treatment in the sheep testis, were rapidly removed leaving later tubules apparently normal except for the almost complete loss of this generation of cells. Hilscher (1964) reported that spermatocytes were the most sensitive cells to lack of oxygen induced by testicular arterial occlusion. Diplotene spermatocytes were considerably reduced in the pig testis.

Dividing spermatocytes are also damaged by heat. The pycnotic and abnormal meiotic figures observed in sheep were presumed to account for the loss of round spermatids observed at a later stage. Heat is thus yet another of the many agents known to cause losses in meiosis. Such losses occur naturally in the reduction divisions of several species, and it has been suggested that this may be a way of excluding chromosomal abnormalities.

By prelabeling cells with tritiated thymidine, it was confirmed that the rate of development of surviving spermatocytes was unchanged by the heat treatment. This supports the concept that the rate of spermatogenesis in mammals is constant and unchanged by variations of daylight, hypophysectomy and subsequent hormone treatment, and the administration of drug that inhibit spermatogenesis.

Spermatids

The early stages of spermiogenesis are affected adversely by heat and the multinucleated "giant" cells often seen in semen after temperature elevation are now believed to be formed by the fusion of damaged spermatids. Immediate losses among the round spermatids were observed soon after temperature elevation in rat, sheep, and pig and similar losses wee inferred from changes in bull semen. Elongating spermatids were probably not affected to the same extent.

The incorporation of lysine into protein by spermatids was also stimulated by glucose, which was believed to offer a partial explanation for the damage to spermatids during insulin hypoglycemia. Temperature elevation may reduce the normal intracellular glucose content in the testis to bring about the same effect.

Spermatozoa

Morphologically abnormal spermatozoa appear in the semen of sheep within 2 weeks of heating, sometimes in a shorter time than that taken by the sperm to traverse the epididymis. Testicular spermatozoa were the first cells to degenerate in mouse testes made cryptorchid, and these spermatozoa showed metabolic abnormalities during the first few days after heating the testes of rams to 40⁰C for 3 hr. Elevated temperature may therefore damage spermatozoa during their later stages of development in the testis or in the early part of the epididymis. Ejaculated spermatozoa of rabbits, if incubated at 40⁰C or made to spend time in the reproductive tract of heated does, are able to fertilize but the resulting embryos are less likely to survive.

Physiology

A number of physiological effects have already been alluded to in the discussion of the effects of heat application to the testis on sperm production and fertility and on the cytology of the testes. Some of the physiological factors considered here include deep testis temperature, blood flow, lymph and tubular fluid flow, and blood gases. The physiological conditions that are altered due to elevated temperature may be of primary concern in determining the mechanism responsible for the degenerative effects.

Deep Testis Temperature

A number of the studies of the effects of seasonal and applied heat to a number of species indicated that there was a change in testicular temperature. However, a number of these studies failed to determine the differential in temperature between the external surface of the scrotum and the deep-testis temperature. Moule and Knapp (1950) designed fine multiple thermocouple which were drawn through the testicles and scrotum of Merino rams. This technique permitted them to record the temperature at various positions in the testis. With the thermocouples in place they subjected rams to control conditions, to hot-room conditions and to a situation in which the scrotum was wrapped in cellophane, cotton wool, worsted fabric, blanketed and then with more cellophane. In control animals the intratesticular temperature varied only slightly. When subjected to the hot room, intratesticular temperatures increased gradually over the $4^1/_2$ hr exposure time. However, the increases did not reach rectal temperature in that period of time. The wrapped testes increased more rapidly in temperature and approached rectal temperature within 3 hr.

Harrison and Harris (1956) demonstrated the effects on rat testicular temperature of applying hot water locally to the testes and scrotum as compared to subjecting the rats to elevated temperature in a hot room. Their results show that there is considerable difference between subjecting the testes to direct immersion in a saline bath or spraying hot water on the scrotum as compared to the change in temperature within the testis in rats placed in a hot room at a temperature comparable to that of the hot water or the saline bath.

Waites and Moule (1961) have also shown the effects of cooling and warming the scrotum on the central and testicular temperatures found in anesthetized rams. They applied direct heat to the testis by means of circulating warm or cool air. Their results showed that if normal heat exchange mechanism are permitted to act, as would be the case with the application of warm air, the intratesticular temperature does not rise to a level corresponding to the temperature of the applied air. This is in contrast to the direct effect of immersing the testes in a saline bath or directly applying hot water, as was carried out in the investigations mentioned above. Thus, while the temperature deep in the testis does tend to remain close to that of the inside of the scrotal skin, the means of applying the elevated temperature is extremely important. Direct application of water at elevated temperature, which would prevent evaporative cooling, results in a rapid change in intratesticular temperature, whereas a slow change usually occur under conditions where environmental temperature is elevated or where the heat is applied by warm air currents.

Blood Flow

Blood flow through the testes might be expected to vary considerably according to the temperature being applied to the scrotum. Moore (1924a) suggested that vascular stagnation with a resulting lack of oxygen and accumulation of carbon dioxide might cause the testicular degeneration that occurred as a result of the application of heat. From the physiological standpoint, however, under most conditions an increase in temperature would be expected to cause an increase in blood flow. Waites and Setchell (1964) found no consistent response in the blood flow after heating the testes of rams to 39^{0}C. Although they obtained a 32% increase in one animal, another three rams were hardly affected. Appelbaum (1964) got no consistent response in the blood flow to the cryptorchid testis of the dog for 2 days following the translocation. However, he did obtain a significant increase with blood flow nearly doubling in the translocated testis compared with the scrotal control after 6 days of artificial cryptorchidism.

Using the isolated perfused testis of the rabbit, Ewing and VanDemark (1963b) reported a decreased rate of blood flow as a result of increasing the temperature to 39.5^{0}C as compared with a temperature of 36^{0}C. Glover and Young (1963) reported that heating produced a characteristic cyanosis in the exposed testis of the rabbit. This was in contrast to the hyperemia that was produced when this treatment was applied to other species. A direct regulatory effect of oxygen on isolated femoral arterial segments from the dog has been noted with an increased pO_2 causing increased vascular resistance and vice versa. This type of local response could explain the decreased blood flow which obtained in rabbit perfusion experiments were 95% oxygen was used for oxygenation of the blood.

Glover (1965) reported that the unilateral cryptorchid rat testis showed a slightly increased blood flow over the contralateral scrotal control. Periods of experimental cryptorchidism were varied from 1 hr to 21 days. The epididymis had a much greater increase in blood flow, particularly in the areas in the body and tail of the epididymis. The effect of a broad range in temperature on blood flow through rat testes was shown using five temperatures between -10^0 and 43^{0}C by Glover (1966). Increases in blood flow were noted in the testes as the temperature was increased, but a much larger increase occurred in the epididymis. Later, Glover (1968) showed that 38^{0}C has a negligible effect on rat testicular blood flow, but that 43^{0}C produced a marked effect. At 43^{0}C the testis temperature was equal to the water temperature after 11 min of immersion. Upon removal from the water bath the blood flow dropped but was not back to normal after 60 min.

Waites *et. al.* (1968) used an isotope-fractionation technique in the rate and found that blood flow through the testis was unchanged at 37^{0}C and only marginally increased at 40^{0}C. However, with this temperature change blood flow through the scrotal skin was doubled. At 43^{0}C blood flow through the testis was doubled and that through the scrotum was approximately trebled. A further increase occurred at 45^{0}C. Blombery and Waites (1966) found that contrary to their earlier findings blood flow in the ram testis appeared to be nonuniform. Using the ^{85}Kr clearance technique they found a fast and slow component of total blood flow which appeared to be representative of 30 and 70% of the testicular mass, respectively. They believe that these components represented the peritubular and intratubular capillary networks. They found similar results with the dog. This finding introduce the possibility that heating brings about a relocation of blood flow without affecting the total blood flow.

Lymph and Tubular Fluid Flow

Another physiological change that occurs in mammals upon elevating the temperature of the testis is the effect on the flow of lymph and tubular fluids. Cowie *et.al.* (1964) found that immersion of the testis of the conscious ram in water at 40^{0}C increased lymph flow rate up to two to three times the control level in 15-30 min and maintained it there until heating was withdrawn. Normal levels were regained within 1-2 hours. The protein content of the lymph (gm/100 ml) did not change appreciably during this time. The increase in flow was due to an increase in the lymph formed because it was maintained over a fairly long period of time. Since the protein content and thus capillary permeability apparently were not altered, this was taken to be a measure of the increase capillary filtration area or capillary filtration pressure (capillary blood flow). Since total blood flow was not greatly affected in this species, it would seem that the capillary filtration area (capillary bed size) must have been increased.

Setchell (1968) showed by measuring weight increase in the testes of rats after efferent duct ligation that fluid secretion was not immediately affected by unilateral cryptorchidism. However, at 4-7 days the secretion rate was less than in the scrotal testis. However, more recently Setchell *et. al.* (1969) indicated that fluid secretion was unrelated to variations in the production of sperm due to scrotal heating. Setchell and Linzell (1968) reported that raising the temperature of the perfused ram testis from 33^0-34^0C to 39^0-40^0C was followed by a fall in retetestis fluid output within 30 min. Thus, the effect of elevated temperature on the secretion of testicular fluid is still in need of further investigation.

Gas Tensions and Levels

Temperature change and the concomitant effects on blood and lymph flow might be expected to change oxygen and carbon dioxide levels in the testis. Most of the studies of these gases have been made by determining the gas tension in the spermatic artery and the spermatic vein or by direct measurements within the testicular tissue. Bergofsky *et. al.* (1962) have suggested that oxygen tension of venous blood is higher than that of lymph which may more closely represent the tissue which it drains.

The blood in the internal spermatic veins is unusually dark, reflecting its relatively low content of oxygen (about 6.5 ml/100 ml blood) and oxygen tension (about 30 mm Hg). Most of this blood comes from the testis but some is from the epididymis and also from fat. Compared with the testis, the epididymis is small (about one-sixth of testis weight in rams), has a higher tissue oxygen tension, and a higher oxygen content in its venous blood. Consequently, the oxygen content of venous blood draining the seminiferous tubules would be lower than that of the venous admixture in the internal spermatic veins, but the difference might be expected to be small. By comparing arterial and spermatic venous blood in the ram, it was shown that the arteriovenous difference of oxygen was 6.2 ml/100 m, giving an oxygen uptake rate of 2.43 µmole/gm fresh wt/hr.

Moore (1924a) proposed that high temperature caused testicular degeneration because of lack of oxygen and accumulation of carbon dioxide "due to vascular stagnation" or "a hyperaemic condition associated usually with edema." Later, Harrison (1956) believed that a locally induced ischemia or anoxia was involved. Both "vascular stagnation" and "ischemia" imply a severe reduction of blood flow, yet Cross and Silver (1962) recorded a rise of 10 to 40% in the pO_2 in the testes of sheep and rabbits during scrotal warming, a result they attributed to vasodilatation and elevated capillary blood flow. The increase was rapid, and the return to normal occurred quickly after cessation of treatment. A moderate rise in pO_2 was also observed with ram testis similarly treated. Since blood flow does not appear to increase with heating in the ram, but oxygen consumption increases markedly, the increase in oxygen tension is surprising. However, a short time lapse was involved in this experiment, so that oxygen consumption may not have had adequate time to become a factor. It is also possible that there is a transient increase in blood flow with heating which was not detected by the investigators cited in the previous section.

Free and VanDemark (1968) found that heating the ram testis to 39.1^{0}C for 2 hr had little or no effect on the gas tension of the spermatic venous blood when measurements were carried out at 37.1^{0}C. If measurements were adjusted to the *in vivo* temperatures a 36% increase in pO_2 from he control to the treated condition was evident. The effects on carbon dioxide were similar. These results indicate an effect of temperature which may be sufficient to account for the rise in tissue pO_2 measured by direct means. Recently, Massie *et. al* (1969) reported that testis oxygen tension (interstitial pO_2) was double the control value in the rat testis translocated to the body cavity for periods of 1, 2, and 4 days. These results may be due to increased blood flow or may be accountable for in terms of a drop in oxygen consumption due to the metabolic changes in, or loss of, temperature sensitive cells.

Although blood flow in the cryptorchid testis several days after operation was elevated above the control, testicular blood flow was unchanged by temperature elevation to 40^0-41^{0}C for periods sufficient to damage spermatogenesis. Nevertheless, heating the testes of conscious rams to 40^{0}C for 2 hr resulted in a sharp increase in the uptake of oxygen by the testis. The oxygen content of spermatic venous blood fell to remarkably low values, and this must have been due to the 70% increase in tissue metabolic rate and extraction of oxygen, since blood flow was not altered. The effect was reversible; when the testis was cooled, oxygen values quickly returned to normal. The tissue was therefore threatened with hypoxia, the severity of which cannot be properly judged without actual measurements of oxygen tension inside the tubules.

METABOLISM

A number of metabolic changes have been observed and followed in the testes after the application of heat. Some of the early changes are likely to be involved in the chain of events leading to irreversible damage of the germinal cells, but few measurements have been made prior to the time that histological lesions become evident. Following histological lesions, metabolic changes may reflect any or all of three types of changes. First, metabolic changes in the tissue may reflect qualitative and quantitative changes in cell population. For example, the loss of germinal cells would result in a change from a metabolic pattern reflecting mainly geminal cells to a metabolic pattern reflecting largely nongerminal cells. The second might involve the changes in surviving cells due to loss of associated cells. An example of this may be the accumulation of lipid in Sertoli cells of the aspermatogenic testis. Third, some metabolic changes may come about due to disturbance of the gonadal-pituitary axis. It has often proved difficult to separate these different contributing factors. However, some pertinent information may be gained from studies where metabolic changes have been correlated with morphological and cellular changes. These later metabolic changes are to some extent the indirect effect of heat on the testis and are of value in understanding the variation in metabolic patterns among the different cells types in the testis and the metabolic relationship between the different components of the testis. It is likely that only the early changes will throw light on the mechanism of heat action upon male gametogenesis.

Carbohydrate and Central Metabolism

Early changes in carbohydrate metabolism of the heat-treated testis have been studied by application of heat to surviving cells *in vitro* or by short-term exposure of the testis to local heat or the abdominal environment. In rabbit testis tissue *in vitro,* an increase in temperature from 33^0 to 37^{0}C resulted in the expected increase in metabolism as measured by oxygen and glucose uptake, and by lactate and CO_2 production. In the isolated perfused rabbit testis, a temperature increase from 36.5^0 to 39.5^{0}C resulted in a decrease of glucose uptake. However, this latter study has been criticized from the point of view that the fall in glucose uptake after 2 hr of higher temperature was attributable to the fall in flow of perfusate and that *in vivo* measurements in the ram testis indicated no such fall in blood flow due to heat Levier (1968) found that the oxidation of glucose and pyruvate by rat testis tissue *in vitro* increased linearly with an increase in temperature from 28^0 to 36^0C. *In vitro* metabolism of rabbit tissue under an atmosphere of air after short periods of cryptorchidism indicated a transitory rise in glycolytic and respiratory activity at 6 hr followed by a sharp fall. A trail carried out in 100% oxygen by these authors did not support this conclusion, however. Furthermore, glucose did not stimulate QO_2 further in control tissues under air but did do so under oxygen indicating that in the former system oxygen was limiting. However, the tissue substrate measurements made in this study 24 hr after translocating the testis showed 12% less glucose and 27% less lactic acid than in control tissues, indicating an increased rate of substrate utilization or a decreased substrate. Other measurements of substrate levels in the cryptorchid rabbit testis indicated no change in tissue levels of glycogen and glucose after 24-hr translocation, whereas lactic acid levels doubled during this period. At 48 hr, however, glycogen was significantly lowered and glucose and lactate were also lower than controls. The depression of substrate occurred somewhat later here than in the previous study. It is also noteworthy that the depression in glycogen levels at 48 hr was also observed in the contralateral scrotal testis of these unilateral cryptorchid rabbits, suggesting that this metabolic change was a result of a disturbance in the gonadal-pituitary axis.

In vivo measurements of arterial venous differences in the ram testis after 2-$2^1/_2$ hours of local heating above 39^0C indicated a 67% increase in oxygen uptake, variable glucose uptake and little change in lactate production. Hexokinase activity of the rat testis was significantly lowered by 24 hr of cryptorchidism (Schanbacher and Ewing, 1969). Phosphofructokinase was also lowered significantly by 48 hr, whereas glucose-6-phosphate and lactate dehydrogenases, and pyruvate kinase were not altered during this period. In support of these data visual estimations from histochemically stained rat testis after 48 hr of cryptorchidism indicated to substantial changes in the activity of succinate, lactate, β-hydroxybutyrate and glucose-6-phosphate dehydrogenases or of NADP and NAD diaphorases. In addition, content and distribution of hyaluronidase, succinic dehydrogenase, or acid or alkaline phosphatase remained unchanged 24 hr after local heat (44^0C for 20 min) application to rat testis. Ford and Huggins (1963) found that levels of malic dehydrogenase as well as lactate dehydrogenase were unchanged up to 4 days of unilateral cryptorchidism in rat.

Some of these studies suggest the possible involvement of a substrate or oxygen shortage in the mechanism of heat effect on the testis. VanDemark and co-workers (1968) have tested the former possibility and found that heat necrosis of rabbit testicle was not prevented by hyperglycemia which elevated testis glucose and glycogen levels. However, in view of the early effect of heat on testicular hexokinase activity, it is possible that phosphorylation of glucose is limiting and therefore that this substrate is not available to the tissue in a normal way. The possibility that hypoxia may be involved in the mechanism of heat degeneration of the testicle has not been tested directly. Measurement of gas tensions in the spermatic venous blood of locally heated ram testis indicate that phyoxia due to increased oxygen demand is not severe. However, there may be a critical threshold for oxygen in the mammalian testis.

Longer term metabolic changes associated with loss of germinal cells due to heat have been discussed elsewhere in this series on the testis. In general it has been shown that glucose loses the ability to stimulate oxygen uptake in the heat treated testicle and may even inhibit it. The respiratory quotient dropped from 0.93 in control rats to 0.52 in the 23- to 28 day cryptorchid. After 2 days of cryptorchidism in the rabbit, tissue levels of glucose and lactate were found to be increased.

Changes in glucose metabolism of rat testis after translocation to the body cavity have been followed for up to 44 days. The fall in testis weight was accompanied by a parallel fall in glucose-^{14}C oxidation to $^{14}CO_2$. Also, a transient fall a gradual rise in recovery of $^{14}CO_2$ from glucose-1-^{14}C and gluconate-1-^{14}C indicated a concomitant change in pentose cycle activity. Glucose uptake also fell in parallel with the change in testis weight, while lactate production remained relatively unchanged. All these changes appear to be associated with loss of germinal cells and the emergence of a pattern of metabolism reflecting that of the surviving cells, i.e., interstitial and Sertoli cells. Apart from these quantitative changes in glucose metabolism some qualitative changes were also observed. Pentose cycle activity as indicated by glucose-1-^{14}C and gluconate-1-^{14}C oxidation was further increased while glucose-6-^{14}C recovery increased relative to the other carbons of glucose and resulted in a recovery rate pattern consistent with the carboxylation of pyruvate and subsequent oxidation by the citric acid cycle. These changes did not appear to reflect loss of germinal cells and occurred too late to be a direct effect of body temperature on the surviving cells. They may have reflected endocrine changes or changes associated with the loss of supporting or dependent cells.

The incorporation of labeled glucose into perchloric acid soluble metabolites was also lower and had a different pattern in the cryptorchid rat testis compared with normal. In contrast to the depression of the oxidation of glucose-^{14}C seen in the rat testis, oxidation of this metabolite to $^{14}CO_2$ was not reduced in the 30-day bilateral cryptorchid rabbit compared with controls. This may have indicated that heat-stable cells also utilized glucose in the rabbit, whereas these cells utilized an alternative substrate in the rat.

In the cryptorchid testis of the rat, the levels of tissue ascorbic acid fell parallel with the loss of germinal cells, reaching a minimum at 15 days. These levels than return to normal by 30 days. As this occurred to a smaller extent in contralateral scrotal testes, it may have represented an indirect change due to disturbance of the gonadal-pituitary axis. Glucuronate-6-^{14}C oxidation remained relatively unchanged in the cryptorchid rat testis. However, myoinositol synthesis from glucose was inhibited in the 7-day cryptorchid rat testis, although the enzyme levels remained unaltered.

Measurements of enzyme activities in heat-treated testes indicated a fall in hyaluronidase activity correlated with the loss of germinal cells, indicating that this enzyme is associated with germinal cells. Similarly, local heat treatment of the rat testis did not affect hyaluronidase activity by 24 hr, but by 21 days hyaluronidase had fallen to a small percentage of control. Other enzyme changes noted were a rapid increase in malic dehydrogenase and lactic dehydrogenase between 4 to 6 days after cryptorchidism in the rat testis and the subsequent maintenance of these elevated levels for up to 74 days. In guinea pig testis, lactate dehydrogenase increased significantly at 3-4 weeks. This increase was mainly due to a change in the muscle-type isoenzyme, whereas the unique testis type isoenzyme fell rapidly during this time and by 5-7 weeks had disappeared. A similar change in testis type lactic dehydrogenase was found in the mouse after unilateral translocation. Sorbitol dehydrogenase was also found to reflect the depletion of germinal cells in the cryptorchid guinea pig testis Glucose-6-phosphate dehydrogenase increased 3-8 weeks after cryptorchidism in the guinea pig. Histochemical evidence in rat testis indicates that this increase occurred largely in the tubular portion of the testis.

Despite the catastrophic effect of heat on the germinal component of the testis, the interstitial tissue was found to be still responsive to gonadotropins and to testosterone propionate depression of oxygen uptake 23-28 days after translocation in the rat. Some histochemical changes have been noted in the interstitium after heat treatment; however, especially notable was a gradual increase in succinic dehydrogenase activity in rat testis.

Lipid Metabolism

Incorporation of palmitate-^{14}C into rat testis *in vitro* was found to have a positive temperature coefficient over an 8^{0}C temperature range. Incorporation into most fractions were increased although monoglycerides and sterols were exceptions to this. The oxidation of palmitate to CO_2 was also increased over this temperature range. Twenty-four hours after local heating (45^{0}C for 24 min) the phospholipid concentration of the rat testis fell. Other changes within the first 48 hr of confinement of the testis in the unilateral cryptorchid rabbit included an increase in concentration of esterified cholesterol and an increase in the level of carboxyl esters and triglycerides. These latter changes occurred in both the translocated and the contralateral scrotal testis indicating a possible endocrine involvement. Phospholipids remained unchanged at 2 days after translocation indicating no loss of cells at this time. Weight both wet and dry, was also unchanged at this time. Although phospholipid level was unaffected 48 hr following translocation of the rabbit testes to the body cavity, the content of palmitate in this fraction increased significantly in both control and translocated testis. Linoleate fell in both testes during this time while the oleate content fell in the translocated testis only. In the triglyceride fraction palmitate content increased in cryptorchid testis by 48 hr.

It appears that the early effect of temperature may include an increased synthesis of glycerides and possibly an increased catabolism of lipids. Analysis of heat-treated testes indicated no effects on phospholipid levels until cell loss occurs. This component is probably related to cell structure. Other changes, including qualitative changes in composition of lipid components may reflect complicating endocrine changes in the testis, especially as many of the changes have been observed in the contralateral scrotal as well as the cryptorchid testis.

Lateral changes in testicular lipids due to heat have been discussed in the chapter concerned with testicular lipids. Most lipid components increase in concentration in the degenerating testis. These changes may reflect a difference in composition between heat-sensitive and heat-resistant cells of the testis. Local and endocrine changes may also be responsible for these increases, as for example, the histochemically detectable accumulation of lipid material in the Sertoli cells and the concurrent changes that take place in the contralateral scrotal testis of the unilateral cryptorchid. As would be expected, total levels of most components fall as germinal cells are lost from the testis. An important exception of this, however, is esterified cholesterol, which may increase or remain unchanged. According to work with heat-treated ram testes this maintenance of cholesterol ester level is due to an increased *de novo* synthesis together with an increased esterification of free cholesterol which more than offsets an increase in catabolism.

Some of these lipid changes due to heat are undoubtedly the result of changes per se in the surviving cells of the testis as concentrations of lipase and esterase in the Leydig cells were increased 4 weeks after translocation in the rat. Similar changes in the contralateral scrotal testis of the unilateral cryptorchid rat suggest that these enzyme changes were a response to increased gonadotropin levels from the anterior pituitary.

Protein Metabolism

In the rat testis, translocated to the body cavity, protein content remained unchanged up to 24 hr but fell significantly by 48 hr. Incorporation of labeled arginine into a number of testicular protein fractions was also reduced within 48 hr of subjection to body temperature in the rat. The effects of temperature *in vitro* on the incorporation of labeled lysine into testicular protein have shown that optimal temperatures were close to scrotal temperatures for rat, rabbit, and mouse, hamster, guinea pig, and dog. Other body tissues had temperature optima close to body temperature. This temperature-sensitive process appears to be specific to spermatids, for in cryptorchid testes where spermatids are lost, temperature optima of lysine incorporation was at body temperature.

The incorporation of glucose-^{14}C into protein of rat testis slices has also been found to exhibit a thermal optimum at 32^{0}C, the approximate scrotal temperature of that species. It appears that an immediate effect of above scrotal temperatures is to reduce protein synthesis in spermatids. However, in the presence of an adequate supply of glucose, *in vitro* incorporation of lysine into testicular proteins at body temperature was similar to that at scrotal temperature. It is possible that higher temperatures in the testis do result in a lowered glucose availability, in which case protein synthesis may be impaired under these conditions.

After a longer period of heat treatment as in the 30-day cryptorchid rat, lysine-^{14}C incorporation into testicular proteins was considerably increased. Incorporation of labeled arginine into total testicular protein was also increased in the 18-day cryptorchid rat. This increase did not occur in all protein fractions, however, since incorporation into nuclear acid-soluble and residual protein was reduced by 80 and 70%, respectively, after 18 days of cryptorchidism. Radioautographic studies have indicated that the rise in protein synthesis in the cryptorchid testis is primarily due to a heat-induced increase in protein synthesis of the surviving cells, particularly of Sertoli cells.

Nucleic Acid Metabolism

No changes in RNA, DNA, or RNA:DNA were observed during the first 48 hr of cryptorchidism in the rat. However, incorporation of cytidine-^{3}H, $H_3{}^{32}PO_4$, and orotic-6-^{14}C acid into total testicular RNA *in vitro* varied inversely with the temperature of incubation between 28 and 32°C. This author suggested that the negative response may involve a direct effect of temperature on RNA synthesis or a secondary effect mediated through thermal activation of testicular ribonucleases. This effect of heat in the testis would seem to deserve a considerable amount of attention in view of the importance of RNA in cellular metabolism.

After longer periods of exposure to heat as in the 30-day cryptorchid rat, incorporation of labeled ATP into DNA by extracts of abdominal testis (expressed in terms of unit quantity of protein in the extracts) was about one-half of that in the scrotal testis. In contrast to this finding, incorporation of thymidine-2-^{14}C into DNA was 1.5 times higher in the 30-day cryptorchid compared with normal mature testes. This apparent contradiction in results between the two studies could be due to different tissue pool sizes of ATP and thymidine. In effect, if pool size of thymidine decreased with cryptorchidism, then labeled thymidine would be less diluted and more label would be incorporated at the same level of DNA synthesis. Incorporation of uridine-2-^{14}C into RNA was increased by a factor of 2.3 in the 30-day cryptorchid testis, but was also increased by a factor of 2.0 in the contralateral scrotal testis.

Concentrations of DNA and RNA in the 30-day cryptorchid testis were similar to normal mature testes. Similarly, with 80- and 200-day cryptorchid rats and with 8-week cryptorchid testis from guinea pigs no change in concentration of DNA and RNA was apparent. As would be expected, total levels of these components diminished progressively with the loss of cells. These results raise the question of why DNA concentration does not increase with the loss of the haploid cells of the testis in cryptorchidism since the remaining cells are all diploid. Hypophysectomy and estrogen treatment do result in a doubling of DNA concentration, whereas x-ray increases it somewhat more than cryptorchidism. It may be that the heat resistant cells, particularly interstitial cells, have more nonnuclear material than the heat-sensitive cells. Furthermore, long-term heat may eliminated the cell stages prior to meiosis, at which DNA is tetraploid and which are still present in hypophysectomized animals. DNA as well as RNA synthesis has been detected by autoradiographical techniques in the germinal epithelium of cryptorchid testis up to 108 days after translocation to the body cavity. It is apparent that considerably more work must be done to clarify these aspects of the heat effect on the testis also.

Endocrinology

Although the effects of elevated temperature on spermatogenesis have long been known, only more recent investigations have helped to indicate some of the effects of temperatures above scrotal level on the hormone balance in the male.

Accessory Sex Gland Changes

As early as 1903, Bouin and Ancel indicated that cryptorchidism had no effect on the accessory sex glands. Others agreed that artificial or natural cryptorchidism produced no apparent effect on the secretion of testicular hormones; thus, suggesting that the Leydig cells were not subject to the heat effect as was the germinal epithelium. Even more recently, Casady *et. al.*, (1953) found no diminution in the libido of bulls, and several investigators found essentially no change in accessory gland weights during cryptorchidism in mice, in the dog, in the rat, and in the ram, bull, and stallion. In fact, Huseby *et.al.* (1961) reported that in mice the weights of the seminal vesicles and coagulating glands increased in the cryptorchid and were maintained at this size of at least a year.

Along with the reported lack of change in the accessory glands have been reports of no change in the interstitial tissue of the cryptorchid testis of the dog and man. Antliff and Young (1957) also found that bilateral cryptorchidism in the guinea pig, starting 2 days after birth, did not interfere with the development or maintenance of the secondary sexual characteristics over a period of 3 years.

Contrary of findings suggesting no change, the interstitial cells in cryptorchid mice have been reported to be larger and to contain more cytoplasm and vacuoles within the cytoplasm than are found in the normal mouse. Reduced weights of seminal vesicles and coagulating glands have been reported following 28 days of cryptorchidism in the rat. Similar findings have been reported by others. Morehead and Morgan (1967) also found that the incorporation of thymidine-^{3}H in the epithelium of the accessory sex glands was significantly lowered in the cryptorchid rat compared to controls. Thus, these changes indicate an alteration in the level of androgens due to a change in the Leydig cells within the testis of the cryptorchid, at least in some species.

Seminal Composition

Engberg (1949) was one of the early reporters of a chemical change in the semen of the cryptorchid. He found the acid phosphatase in the semen of cryptorchid men was lower than normal. Fructose level of the coagulating gland and the seminal vesicle output of critic acid were found to be altered in the artificially cryptorchid rat with below normal levels at 5 days, increased levels between 10 and 20 days after the operation, and finally subnormal levels again after the peak period. P. Collins and Lacy (1967) reported normal fructose levels in the seminal vesicles of rats during a 3-day to 6-week period following a local exposure of the scrotum to 43^{0}C for 10-30 min. Although Skinner and Rowson (1967) found no effect on the seminal vesicles or bulbourethral glands of the ram, bull, and stallion, the ampullae on the cryptorchid side in their experimental animals weighed less and had lower fructose and citric acid content. Glover (1956), on the other hand, found that scrotal insulation of rams produced a fairly rapid increase in seminal fructose which gradually returned to normal after discontinuing the treatment. The libido of the rams was unaffected and the seminal fructose level decreased prior to the decline in libido at the end of the breeding season. Although these chemical tests on semen were considered to be more exact indicators of androgen production by the testes, the results were still not as precise as were needed to assess more accurately the androgen output of the testis.

Androgen Levels

The urinary output of steroids by cryptorchid rabbits and men has been found to be considerably reduced from that of normal individuals. The levels usually were somewhere between the normal and the castrate. Engberg found the levels in the urine were correlated with the acid phosphatase in the semen.

One of the early assays of the testis for androgen content was performed by Hanes and Hooker (1937) on the cryptorchid testis of swine. As determined by their bioassay, the cryptorchid testis contained approximately one-half as much androgen per unit of weight as did normal testis. S.A. Gunn *et. al.* (1961) found a 41% decrease in androgen levels in rats as indicated by the zinc uptake of the ventral prostate within 1 week after experimental cryptorchidism. Eik-Nes (1968) found that in the congenital unilateral cryptorchid dog secretion of testosterone into the spermatic vein blood was 30% lower from the cryptorchid testis than from the normal testis. The absolute content, but not the concentration of testicular androgen, was reduced 10 times in the unilateral cryptorchid testis in studies by Skinner and Rowson (1967, 1968) utilizing the ram, bull, an stallion. P. Collins and Lacy (1967) separated the tubules and interstitia of the rat testis for steroid analysis and found no change in the steroid content of the interstitia due to exposure to 43^{0}C local heat to the scrotum for 10-30 min, but there was an increase in the tubular steroids.

Eik-Nes (1965) reviewed the effect of temperature on testosterone production by the dog *in vivo* and *in vitro*. Prolonged (60 min) temperatures of 39.5^{0}C depressed testosterone production. There was only slight variation in the *in vivo* secretion of testosterone by dog testis between 35.5 and 39.7^{0}C. The peak occurred at 38^{0}C, and at temperatures below scrotal temperature testosterone synthesis fell considerably with a rapid recovery on rewarming. Inano and Tamaoki (1968) found that *in vitro* conversion of pregnenolone to testosterone in the rat was lower in the cryptorchid testis. Eik-Nes (1965) also showed that dog testis tissue receiving arterial blood 20% the normal rate had a limited ability to regain its biosynthetic capacity to produce testosterone.

The capabilities of testicular tissue to respond to the stimulus of HCG in producing testosterone as temperature increases up to 38^{0}C, have been shown by Eik-Nes (1965). However, the capability of cryptorchid testis tissue to respond to the stimulus of gonadotropin is considerably less than that of normal tissue. Clegg (1960) suggested that a rise in androgen levels in the initial stages of cryptorchidism was a response to a higher gonadotropin level, and then later the androgen levels diminished as the heat-damaged tissue become incapable of responding to the gonadotropins. Clegg (1961) suggested that the increased number of interstitial and Leydig cells following artificial unilateral cryptorchidism was also a response to the increased gonadotropin levels. A lack of response to gonadotropins by cryptorchid tissues was also shown by the lack of increase in 17-ketosteroid output in the urine of cryptorchid men treated with high levels of exogenous gonadotropin.

The interaction of temperature and gonadotropin level with rat testis tissue was demonstrated by Gospodarowicz and Legualt-Demare (1962). They found that a greater percentage response to HCG occurred at 27⁰C than at 33 and 38⁰C in the incorporation of acetate-1-^{14}C into cholesterol. However, incorporation into androstenedione was depressed (57%) with increased temperature alone up to 38⁰C, whereas an increased incorporation of acetate into cholesterol occurred with higher temperatures.

The ability of the cryptorchid testis of the rat to convert progesterone to 17a-hydroxyprogesterone, androstenedione and testosterone was significantly lower than the capabilities of scrotal testis tissue from rats. They also found that the steroidogenic response to HCG was less in cryptorchid tissue. Also working with the rat, LeVier and Spaziani (1968) found that the *in vitro* incorporation of labeled cholesterol into components of the Δ^4 and Δ^5 steroidogenic pathways had a thermal optimum of 32⁰C. Hall (1965) found that incorporation of labeled acetate into testosterone by rabbit testis tissue slices *in vitro* had a maximum rate at 38⁰C, which corresponded to scrotal temperatures in this rabbits, with a sharp decline at higher temperatures. Androgen production from pregnenolone *in vitro* also was lower in rat cryptorchid testis tissue. Local heating of the testis in the scrotum to 43⁰C for 30 min in rats did not change the ability of interstitial tissue to incorporate labeled pregnenolone and progesterone into steroid intermediates *in vitro*; tubular tissue incorporation was relatively low but was increased by the previous heat treatment. Gomes and Johnson (1967) found that whole-body exposure of rams to 32⁰C for a 2-week period depressed the capacity of testis tissue slices *in vitro* to synthesize androgens from various precursors.

A reason for the depressed capabilities of tissues in the cryptorchid animal to carry out steroid metabolism is found in the observations that the unilateral cryptorchid testis tissue of the rat lost most of its 3β-hydroxysteroid dehydrogenase activity after 64 days translocation. No change was seen up to 8 days. Others have reported decrease in the activity of this enzyme with cryptorchidism in the rate, ram, bull, and stallion. The reductions of this enzyme were rather marked while certain other enzymes were increased in the cryptorchid testis.

Estrogenic Hormones

With the decrease in androgen secretion which accompanies cryptorchidism it has been suggested that there is an increase in estrogen levels, but evidence for this is scanty. The seasonally cryptorchid testis of the stag has been reported to have four times the amount of estrogen found in bull testes. Takewaki (1957) found that cornification of vaginal grafts required lower doses of estradiol when the grafts were made in cryptorchid rats as opposed to being made in normal controls. This lower requirement for estradiol was interpreted to indicate that a higher level of estrogen was present in the cryptorchid animals, but one could also argue that a lower level of androgen antagonisms was present. Estrogen-producing Leydig cell tumors in dogs have been reported, and estrogen extracted from a Sertoli cell tumor has been found to be twice as high as the amount found in the ovary of an estrous bitch. One can envision the possibility that androgens are converted more readily to estrogens in cases where the temperature is higher. Hill (1937) demonstrated that ovaries grafted into the ears of castrated male mice secreted amounts of male hormones sufficient to maintain normal seminal vesicles and prostates. The male hormone secretion of the grafted ovaries was considerably enhanced by cooling the grafts and their surroundings to 22^{0}C. When the grafts were maintained at near body temperature (33^{0}C) for 156 days, no evidence of male hormone secretion was detected, suggesting that higher temperatures were necessary for the conversion of androgenic precursors to estrogens or that alternate pathways, resulting in more potent androgens, were favoured by the lower temperature. Ovarian grafts did maintain the accessory glands of castrated male mice for 100 days to 1 year.

Gonadotropins

Several investigators have suggested an increase of the gonadotropin level in the cryptorchid animal. This conclusion has been reached by the fact that the accessory gland growth is increased (in the mouse) and by the increased output of fructose and citric acid in the semen during the first 3-week period following artificial cryptorchidism in rat and ram. The anterior pituitary in the cryptorchid rat has been shown to have more basophilic cells at 75 days with vacuolation starting at 140 days and becoming advanced at 200 days. At puberty, more signet rings appeared in the pituitaries of rats made cryptorchid at 4 weeks of age than in normal rats. Marinosci *et.al.* (1957) also demonstrated more pituitary gonadotropin in bilaterally cryptorchid rats during the first month after operation with no significant change thereafter up to 4 months. Even in earlier studies more gonadotropins were reported in the urine of cryptorchid men than in normal men. In their studies of the 17-hydroxysteroid dehydrogenase activity of normal and cryptorchid testes, Llaurado and Dominguez (1963) found that hypophysectomy had a greater depressive effect on the cryptorchid testis than on tissues from the scrotal testis. Likewise, the stimulatory effect of HCG was less in the cryptorchid testis.

The separate roles and the interaction of pituitary and testicular hormones, as they are affected by the temperature of the testis, are still not well understood. The fact that a unilateral artificial cryptorchidism will bring about degeneration of the translocated testis at a more rapid rate if there is a scrotal testis present than is the case if the contralateral testis is removed or if a bilateral cryptorchidism is performed, is highly indicative of this interaction. Whether the hypothesis that bilateral cryptorchidism brings about a reduced release of testosterone and thus triggers a greater release of gonadotropin which in turn sustains the testis at body temperature for a slightly longer period is yet to be proved.

THE EFFECT OF COLD

Although considerable attention has been paid to the effects of heat on testicular function, presumably because of the naturally occurring cryptorchidism in many species and the extreme sensitivity of the testis to heat effects, much less attention has been paid to the effects of low temperatures. This undoubtedly is due to the fact that the temperature controlling mechanisms in scrotal animals permit adjustments which prevent the exposure of the testes to extremely low temperatures. It is also obvious that the functional activities of the testis can be maintained over a much wider range of decreasing temperatures than is the case with increasing temperatures. However, evidence does exist indicating that extremes in cold will affect not only fertility and sperm production but also the histological and physiological conditions of the testicular tissue.

Whole-Body Cooling

Compared with heat, cold poses a less severe threat to spermatogenesis in mammals. For example, male rodents can remain fertile at near zero air temperatures. Mice in a meat cold storage facility at about -15⁰C, made burrows into the meat, which they lined with cloth wrapping off the meat, and remained fertile. However, when deprived of the means of creating a microclimate, rats living at

6°C for 3 months suffered reduced testis weight or total inhibition of spermatogenesis. However, the *in vitro* metabolism of rat testis and spermatogenesis of mice and guinea pigs experiencing similar cold stress was unaffected. Testicular function was impaired within 11 days in male rats living at -5°C. This effect was thought to be mainly the result of reduced androgen secretion as judged by involution of accessory glands. The testes had to be anchored in the scrotum to prevent their reflex withdrawal into the abdomen and the consequences of cryptorchidism.

The sex drive and fertility of rats were reduced after moderate hypothermia, when body temperature was made to fall to between 15 and 20°C, and during which the animals experienced hypoxia and hypercapnia. Male rats further cooled so that the respiration and circulation stopped for 1 hr when body temperature was 0° to 1.5°C, suffered reduced fertility during 8 weeks reanimation. Spermatozoa and spermatids were damaged but in the most tubules spermatogonia and spermatocytes survived.

Direct Cooling of Scrotum and Testis

Locally cooling the rabbit testis to 1°-3°C by applying ice to the scrotum induced decapitation of spermatozoa in the epididymis but did not harm spermatogenesis. Fertilizing capacity of spermatozoa was reduced after cold treatment, and the germinal epithelium was damaged if the local cooling was so severe that testicular temperature fell to -0.8° to 2.5°C. Directly immersing the exposed testes of rats into a bath at 2° and 0°C for 1 hr had little effect on spermatogenesis, whereas cooling the testis to -5°C or lower produced severe destruction. If the scrotum was retained around the testes, spermatogenesis survived immersion at -4°C and little damage occurred until the temperature of the bath was reduced to -10°C. Similar results were obtained with the guinea pig.

Physiological Responses

The effect of temperatures of -10°C in the rat and -13°C in the guinea pig on intratesticular temperature of the testis with and without the protection of the scrotum have been recorded. The reports given by both sets of investigators indicate that intratesticular temperature drop rapidly if the exposed testis is immersed in an ice-salt mixture, and intratesticular temperature drops to near the 0 mark or below in 10 to 20 min in the rat and to even below 0 in less than 10 min in the guinea pig. However, exposure of the scrotum to similar temperatures reduced intratesticular temperature in the testis of the rat to only about 2^{0}C and in the guinea pig only at an exposure of -10^{0}C did intratesticular temperature drop below 0. By using a technique of injecting radiopaque material into the aorta, MacDonald and Harrison (1954) showed that the effect of cold on the testis unprotected by the scrotum was probably caused primarily by ischemia. This was likely due to the vasoconstriction produced by the low temperatures, for the radiopaque material did not progress beyond the point at which the testicular artery curved posteriorly into the testicular parenchyma in the rat at a temperature of -8^{0}C. Each 2^0 higher temperature permitted a slightly further progression of the opaque material into the arterial system of the testis. The materials penetrated deeply into the arterial system of the control testis and epididymal vessels.

Glover (1966) showed that the testis suffered the greatest decrease in blood flow due to temperatures of 0 and -10^{0}C compared to the epididymis, tunica vaginalis, and scrotum in rats. These results are similar to the finding reported by Cross and Silver (1962) on changes in the oxygen tension induced by temperature variation. Thus, the effects of low temperature on the testis may be primarily the effect produced by ischemia.

Perrault and Dugal (1962) demonstrated that the effect of cold has a specific depressive effect on the weight of the accessory glands in the rat, and they considered this as indicating a decreased androgenic function due to the cold alone. Several examples of decreased metabolic activity by testicular tissue due to cold have been cited by others in this volume.

4

Biological Effects of Low Temperatures

The degree of the cell injury caused by low temperatures depends on the freezing protocol (the cooling rate, the thawing rate, the exposition duration, the minimal temperature) as well as on the cell environment (an isolated cell in the solution, a densily packed ensemble of the cells in the tissue sample–in vitro; the cells in a living organism–in vivo).

Processes in Cells Under Hypothermia

The adaptive reaction of the living organisms to cold includes three components:

1. To perceive (detect) low temperature
2. To transduce the signal (probably, by initiating the signaling cascade) to activate or repress expression of appropriate genes
3. To utilize the genes to start the necessary changes of biochemical processes

Some species not only survive but remain active at temperatures near 0 °C or lower such as snow fleas (Collembola) that could be seen hopping over the surfaces of glaciers. It is hoped that understanding cold tolerance mechanisms can help in the development of the freeze resistant crop plants by gene transfer. It is, however, not a simple task, since cold resistance does not controlled by a single gene – it is a complex of the interrelated processes, including regulation of the membrane fluidity, synthesis of low molecular weight, and high molecular weight cryoprotectants. At least 50 genes are identified now in different species that could be sorted into the following groups:

1. Genes that encode structural proteins participating in protection of the cell against low temperature stress
2. Genes encoding enzymes involved in synthesis of different osmoprotectants (including amino acid derived ones), in desaturation of lipid macromolecules, and in the antioxidative response
3. Regulatory genes that control the overall mechanisms of defence against cold stress

Two strategies for the adaptation of the organism to low temperatures are observed in nature:

1. Avoiding the crystallization: an accumulation during the acclimation period of oligosaccharides, other solvable carbohydrates, glycerol, free amino acids (especially proline, glycine, and betaine) that prevent the ice nucleation by increasing the water binding capacity and stabilize the protein structure; the multicomponent nature of the resulting cryoprotective mixture reduces the possible toxic effects associated with the concentration required for the single component to produce the same level of the cryoprotection. Achieving this goal is also assisted by antifreeze proteins, AFPs (thermal hysteresis proteins, THPs), that reduce the crystallization temperature while keeping the melting temperature unaltered; an amplitude of the hysteresis depends on the AFP concentration; lowering of the freezing point by AFPs action is noncolligative nonequilibrium. Sometimes solutes accumulated by the cells at low temperatures are referred to by the unifying name "*compatible solutes*". Cold acclimation is also accompanied by changes in gene expression; some of the corresponding proteins, especially protectin, stabilize cell membranes during freezing, others are engaged in providing anoxia tolerance and antioxidant defence.

The importance of the acclimation period could be illustrated by a couple of examples:

(a) G. Bryant and J. Wolfe observed significant differences in the survival rates (correlated with the morphological changes in the cellular membranes) during freezing for the protoplasts extracted from the leaves of the cold acclimated and unacclimated rye seedlings;

(b) Needles of the Central European Scots pine *Pinus sylvestris* L. are lethally damaged by exposition to the temperature −10 °C during summer while in winter they endure the exposure to −50 °C.

It should be noted, however, that examples of very fast reaction of organism to cooling are also observed in nature: the alpine plant *Lobelia telekki* growing on the slopes of Mount Kenya as able to cope, mainly by INA production, with temperature fluctuations over the range −10 °C to +10 °C during a one day period.

2. The regulation of the crystallization:
   (a) *in time* (to be more exact, in the *temperature range*, e.g., by the minimization of the supercooling ΔT and thus limiting the nucleation rate that strongly depends on the supercooling $J \propto \exp(B\tau)$, where $\tau \propto (\Delta T)^{-2}$, ΔT is the supercooling); following ice nucleation and initial crystal growth the temperature rises to the melting point and remains approximately constant during the subsequent crystallization of water – the so-called "*latent heat plateau*".
   (b) *In space* (ice formation in the extracellular space): modification of the ice formation process by synthesis of ice nucleating agents (INA) – *nucleator* proteins that assist in the formation of nuclei of the critical size at the desired temperature (as a rule, at higher temperature than would occur in their absence, usually ≥ −10 °C) and in those parts of the organism where the presence of ice is not lethal (the extracellular space where the nucleator proteins have been found in the living organisms). Freeze-tolerant animals endure the presence of ice in all extracellular spaces (e.g., in abdominal cavity, bladder, brain vetricles, eye lens) and survive the temporary interruption of all vital functions (breathing, blood circulation, nerve, and muscle activity), regaining them upon thawing.

The largest of freeze-tolerant animals are the box turtles *Terrapene carolina* and *Terrapene orusta* whose mass is about 0.5 kg. Most freeze-tolerant animals are from North America; exceptions are the Siberian salamander *Salamander keyserlingi* and the European common lizard *Lacerta vivipara*. Note, however, that some minor freezing injuries still could occur in surviving freeze-tolerant animals.

Different species "prefer" different cryoprotectors. For example, glucose is the only CPA produced by *R. sylvatica* (the reactions of synthesis of glucose from glycogen by frog liver are the same as those used by all vertebrates), while grey tree frogs and Siberian salamander use glycerol as a primary CPA. Relatively low level of colligative CPAs in reptiles correlate with their generally lower freeze tolerance compared with amphibians. Overwintering insects may contain as much as 25% of glycerol, for example, larvae *Bracon cephi* and *Eurytoma gigantea*.

Glucose and glycerol are each highly efficient in preventing hemolysis of wood frog erythrocytes, while glucose turn out to be a poor CPA for the human red blood cells.

The controlled ice growth allows, inter alia, to provide the dehydration of the important organs and to avoid their injury; direct evidence of the relationship between maintenance of the critical minimum cell volume (CMV) and freezing survival was obtained by Storey, Bischof, and Rubinsky in experiments on freezing in the liver of *Rana sylvatica*. Cell shrinkage during extracellular ice formation is minimized by colligative cryoprotectants. Rapid distribution of cryoprotectant is ensured by a high density of the membrane transport proteins. AFPs are found in the organisms of this group along with the nucleator proteins; their task in these animals is to prevent the recrystallization of ice during thawing. The survival of freeze-tolerant animals also depends on the freezing rate, since the low rate of the temperature change provide enough time for implementation of the cryoprotective measures such as the distribution of the cryoprotectant throughout the body. Control of the nucleation start allows these animals to avoid dangerous fast ice formation that occurs if supercooling is too large. For example, typical crystallization rate in wood frogs is less than 5% of the total body water per hour.

In contrast to freezing, which is a directional process with ice propagating inwards from the periphery to the core of the body, thawing occurs relatively uniform thoughout the body. Internal organs such as heart or liver melt quickly being still surrounded by the extra-organ ice since high concentration of CPA results in a low melting point. In fact, the heat beat is observed to resume before all ice in the body is melted.

Important factors of freeze-tolerance are mechanisms of anoxia tolerance for sustaining cellular metabolism in the frozen state; sometimes cold adaptation involves a conversion from an aerobic to a partially anaerobic metabolism at low temperatures. Oxidative stress at low temperatures in plants is usual due to the reduction of the photosynthesis rate.

The change of the adaptation strategy depending on the state of the environment is also observed; the European common lizard *Lacerta vivipara* could serve as an example of species that use both strategies.

Freeze avoidance is considered to be less favorable than freeze tolerance since the former is associated with significant energy loss.

Although some general rules and similarity across taxa exists, different species, even within the closely related taxonomic groups, have evolved different mechanisms to cope with the same problem of low temperatures. As R. Lundheim once noted: "you can on the same stone or the same tree find lichens with active AFPs, without AFPs and with potent nucleators and all combinations on the same tree".

The ability of plants to endure cold to a great extent depends on the environment. D.J. Beerling et al. studied the effects of the long-term global climate changes (CO_2 enrichment and increased exposure to ultraviolet (UV) radiation) on the ice nucleation temperatures of some evergreen and deciduos woody shrubs and concluded that these changes will lead to the significant increase of the foliage damage of subarctic plants.

Freezing can induce the so-called outgasing due to the variation of the gas solubility in water with temperature. Its best known manifestation is in the most embolism-sensitive biological system: xylem vessels in cold climate freeze and embolize annually. Even a moderate amount of embolism has a significant effect on the xylem water balance: the stem diameter swells and the daily diameter variation increases.

Antifreeze Proteins (AFPs)

AFPs were first described by Ramsay in 1964; later AFPs and antifreeze glycoproteins (AFGPs) were found in the species of different taxonomic groups, including Arctic and Antarctic fishes such as Antarctic cod, Antarctic and Arctic eel pout, Arctic polar cod, Winter flounder, over 40 species of insects (mainly beetles), bacteria, fungi, and plants including grass. Antifreeze proteins perform dual function:

1. Preventing freezing from external ice
2. Inhibiting internal ice growth

The important action for fish is suppression of ice growth, not of nucleation: some fishes were found to contain ice crystals that do not grow even under supercooling.

Insect AFPs are considerably more active than those from fishes; the *thermal hysteresis activity* (THA) of the plant AFPs is low compared to that of fishes and insects (the levels of THA for plant AFPs are in the range 0.2–0.6 °C), their inhibition of the ice recrystallization is, however, comparable or greater. There is a simple explanation of this fact: supercoolings involved in recrystallization are very small and thus a moderate hysteresis activity is sufficient to suppress the process.

The different activity of the fish and insect AFPs is attributed not to the different affinity of AFP to ice, but to the different morphology of the ice crystals that grow in the presence of AFP: all fish AFPs shape ice into a hexagonal bipyramid that bursts out of tips along the *c*-axis, while AFP from the spruce budworm *Choristoneura fumiferana* (CfAFP) having roughly four times the thermal hysteresis of type I fish AFP produces ice crystal that bursts along *a*-axis. THA is the effect that is easiest to quantify and thus it is usually used to characterize the strength of AFPs. While the THA of fish AFPs (approximately 0.7–1.5 °C) is considerably lower than that of AFPs from terrestrial arthropods (3–6 °C), this level is sufficient to depress the freezing point of the body fluids below that of seawater, thus protecting freeze-avoiding Arctic and Antarctic fishes. The measured THA of certain insect AFPs is shown to inversely depend on the size of the ice crystal present in the sample.

Fish AFPs are the most characterized and usually are divided into several groups according to their structure.

The first one contains AFGPs and type I AFPs containing linear helical structure with periodically repeating ice-binding elements. Type I AFPs are alanine-rich á-helical proteins with the molecular mass ca. 3,300–4,500Da.

The second group comprises globular proteins without any apparent repetitive substructures; it includes type II (cysteine-rich globular proteins containing five disulfide groups) and type III (globular proteins having the molecular mass about 6,000 Da) AFPs.

The more lately discovered type IV glutamate- and glutamine-rich protein is considered as being intermediate to these groups and is assumed to have an á-helix bundle structure. The AFP molecules, as could be seen, are rather large; the smallest one discovered so far has the molecular weight about 1,600.

In fact, two different AFPs were found in fish. The first one was largely confined to the liver and secreted into the blood to be distributed throughout the extracellular spaces of the organism. The second, *skin-type* AFP, is expressed in a wide range of tissues, including gills, most abundantly in external epitheliathe tissue that come into the intimate contact with ice-laden environment.

Antifreeze glycoproteins have a molecular weight from 2,600 to 34,000, while the antifreeze proteins between 3,200 and 14,000. In the solution these compounds usually form helical rods. They are amphipathic molecules, with one side of the rod being composed of hydrophobic residues and the other side being composed of hydrophilic residues. On the hydrophilic side, there is a 4.5Å separation between repeating threonine and aspartate residues that bind the protein to an ice lattice.

There are great differences in both the primary sequences of antifreeze proteins and their 3D structure. Still, there are some common features such as shaping of both the fish and insect AFPs in such a way that the binding sites through which AFP can dock to ice are relatively flat (the area of the binding surface is about a few hundred square angstroms) and constitute a significant part of the protein surface. It was suggested that the threonine and aspartic acid residues could form hydrogen bonds to oxygen atoms on the primary prism plane of ice.

There are a number of contributions to the energy of binding, in addition to the hydrogen bonds, namely, enthalpic contributions from the van der Waals forces that come from the ideal surface–surface complimentary and probably the entropic contribution that follows from the fact that there is no need to solvate the hydrophobic ice-binding surface when it docks to ice.

THA of AFPs depends, in addition to their specific activity and concentration, on the presence of the so-called *enhancers* that may be low-molecular-mass substances such as glycerol and citrate or other proteins. For example, other AFPs or thuamatine-like proteins could serve as antifreeze protein enhancers forming complexes that block a larger surface area of the potential seed ice crystal. The reported enhancement of THA by the low-molecular-mass solutes varies from threefold (glycerol, sorbotol, alanine, ammonium bicarbonate) and fourfold (succinate, aspartate, glutamate) up to sixfold (citrate). These enhancers could also inhibit action of INAs.

Genes encoding AFPs cloned from some insects that are among the most active known natural AFPs, which are considered for introduction into plants to increase their cold resistance.

In contrast to cryoprotectants such as glycerol, sugars, and DMSO that act colligatively and thus need to be at rather high concentration to produce the required effect, AFPs' action is of non-colligative nature and therefore are preferable for the use in food industry since extremely low concentrations of these substances (hundreds times lower than corresponding colligative CPAs) will hardly have adverse effects on the taste, the texture, or the toxicity of frozen food products. S.R. Payne et al. reported beneficial effects of the meat treatment with

AFGPs before freezing: the samples after thawing exhibit less drip with no detrimental effect on texture, tenderness, or juiciness. Addition of AFPs also reduces the microbial activity within food products.

Ice Nucleating Agents (INAs)

The first INA was found in bacteria isolated from frost-damaged plants; it is assumed that these nucleators induce the damage to the host thus giving the bacteria access to nutrients from the plant. Ice nucleators have been found in many other organisms, including lichen, fungi, insects, mollusks, and vertebrates. INAs are thought to be active when they are attached to the surface of cellular membrane and to keep their properties (ice-nucleation activity) even for 1:1,000,000 dilutions. Extracellular INAs actually prevent the formation of intracellular ice by initiating crystallization in the extracellular space at high sub-zero temperatures.

In plants, both extrinsic INAs such as ice-nucleating bacteria (e.g., *Pseudomonas syringae*) and intrinsic nucleating agents synthesized by plants are acting. These nucleators induce the stems to freeze at warm, subzero temperatures. The ice propagation into the lateral appendages such as buds or newly extended primary tissues is prevented by hydrophobic barriers that are most effective if the initial freezing occurs at the relatively warm temperature.

A new class of heterogeneous ice nucleators was found in the freeze-tolerant larvae of the gall fly *Eurosta solidaginis* - probably, the most popular insect in cryobiology. Large crystalloid spheres present within the Malpighian tubules of the larvae turned out to be a hydrate of tribasic calcium phosphate. Spherule size usually varies from 100 to 300 μm in diameter, some large spherules over 500 μm were also observed. Unlike many known heterogeneous nucleators that have a hexagonal physical structure similar to that of ice lattice (such as some organic compounds or silver iodide well known to atmospheric scientists), amorphous calcium phosphate spherules could induce ice nucleation via cracks or other structural defects. Body fat cells seem to be another source of ice nucleators.

Ice-nucleating activity of spherules is less than that of ice-nucleating proteins and lipoproteins. Still, their activity, along with the contribution from the body fat cells, provides a sufficiently high shift of the whole-body nucleation temperature needed for the survival of *E. solidaginis*.

Interaction of AFP or AFGP with ice surface is a specific example of the general problem of protein interaction with solid surfaces that is of great interest for bio- and nanotechnology (integration of an implanted device or material with tissue, hybrid biological/inorganic devices, bio-nano-assembly technologies).

The most important distinction of the freeze-tolerant organisms is the difference between the freezing temperature and the lethal temperature.

Often the cold tolerance of the animal is expressed in terms of the freezing temperature (supercooling point (SCP)) alone. Bale notes that the greatest threat to survival is the cumulative effect of the prolong exposure to low temperatures rather than freezing per se and thus the dichotomy "freeze-tolerant–freeze-avoiding" is too coarse to describe the variety of the organism's reaction to cold observed in nature. Bale proposed to extend the classification by adding three more groups of organisms:

1. *Chill tolerant*: species with low SCP (usually from –20 °C to –30 °C) that differ from the freeze-avoiding group by occurrence of some mortality above the SCP that is increasing with the exposure time
2. *Chill susceptible*: species that die after brief exposures (minutes or hours) to temperatures significantly higher than the SCP
3. *Opportunistic survival*: species that are unable to survive below the threshold temperature for development

The lowering of the temperature of the organism initiates the expression of the heat shock proteins (HSP)[2] (also called *heat stress proteins* or *cold shock proteins*) that assist in keeping the cell homeostasis (e.g., by compensation of the reduction of the metabolic reaction – the temperature effect is described by the Arrhenius equation $k = Ae^{-Ea/RT}$, that is the exponential decrease of the reaction rate with the temperature; frequently the mediators are the cold adaptive, or psychrophillic enzymes, both intracellular and extracellular, that are distinguished by the reduced activation energy E_a and enhanced catalytic activity).

The initial stage of adaptation to cold is the detection of low temperatures by the organisms and the transduction of the signals to activate the biochemical changes, including the gene response that provide defense against cold. The plasma membrane is considered as the principal sensor of low temperatures since the membrane rigidification causes the cytoskeleton rearrangement and changes in the ions influx to the cell. The direct proof of the effect of the membrane properties on the gene expression was obtained by Orvar et al. in experiments on the fluidization of the membrane with benzyl alcohol and the membrane rigidification with dimethylsolfoxide. The membrane rigidification could, in turn, trigger the reorganization of the cytoskeleton. In addition to the direct effect on the membrane fluidity, low temperatures influence stability of the secondary structures of ribonucleic acid (RNA) and DNA as well as the activity of enzymes, including those involved in the transcription and the translation as well as the metabolic reactions.

Studies of the signal transduction in *Arabidopis thaliana* and other plants showed that intracellular calcium plays a significant role in the process of cold adaptation: all cell types respond to cold by elevating calcium (probably, by a direct activation of the mechano-sensitive calcium channels via the changes in the mechanical state of the membranes); in fact, the elevation of calcium could be caused by different abiotic stresses, including salinity. Increase of membrane viscosity affects Ca^{2+} channels also, causing an increase of the cytosolic calcium concentration.

A number of components are considered as the probable intermediate agents in the transduction of the calcium signal: calmodulin, calcium-sensitive protein phosphotase, calcium-dependent protein kinase.

The increase of the intracellular concentration of calcium, in turn, activates expression of the transcription factors that control the subsequent transcription of several cold-regulated genes.

Many genes are regulated by different biotic and abiotic factors (one speaks about *cross-talk* between different stress signaling pathways in this case), but some of them are specifically regulated by low temperatures.

Cold sensing in plants also could be related to the temperature effects on the metabolic reactions and imbalances between the absorption of light energy and its metabolic consumption.

The observations indicate that cells detect the rate of the temperature change d*T*/d*t* rather than the temperature itself. More detailed information on the molecular and gene processes of the cold adaptation of the psychrophilic organisms could be found.

Notwithstanding the aforesaid, the cooling could result in the cell death: the tertiary protein structure as well as the polypeptides contained in the protein are changed and the denaturation of the latter is possible. Cooling could also cause the mechanical damage to the cytoskeleton, the structure of which is determined by the bonds between the membrane proteins and the cell scaffold, since lowering the temperature weakens these bonds. Cooling can cause marked effect on microtubules, including depolymerization and disappearance of the microtubule organizing centers. Such behavior is observed in unfertilized oocytes. Low temperatures can also act on other main components of the cytoskeleton, for example, on microfilaments composed of polymerized actin (F-actin) that are important for different cell processes, including cytokinesis. The action of low temperatures on the cytoskeleton could be mediated by increased level of the intracellular calcium that activate calcium-dependent proteases and phospholipoidases.

Hypothermic damage of cells was extensively studied with respect to the long-term RBC storage and even a special term *hypothermic storage lesion* was introduced. It could include adenosine triphosphate (ATP) depletion, which is critical for the maintenance of electrolyte balance and powering sodium and potassium ionic pumps as well as membrane injury that is observed as lipid loss, microvisiculation, macroaggregate formation, and reduction of the ratio of the surface area to the volume.

Changes of the cell state caused by low temperatures could manifest themselves in the behavior of the larger biological structures comprising these cells. Thus, elevation of the intracellular Ca^{2+} in cardiomyocytes contribute to the posthypothermic myocardial failure.

Protein Denaturation

The stability of the native protein conformation is determined by the difference of the Gibbs free energy of the native and denaturated states, which is a parabolic function of the temperature with a maximum in the range 15–25 °C; a hydrogen bonds net in the system of the water molecules interacting with the protein plays a significant role in the process of denaturation; for an additional information on the protein stability please consult a review by Rose and Wolfenden.

Cold protein denaturation (a distortion of the secondary structure defined mainly by hydrogen bonds, although some disulfide bonding can also occur), observed in the temperature range 0–20 °C could be either reversible or irreversible in contrast to the usually irreversible thermal denaturation; in particularity, the freezing of the globular proteins frequently results in the reversible partial denaturation. During denaturation the individual covalent bonds between the atoms of the polypeptide of the protein molecule are not broken; just rearrangement of the hydrogen bonding from the ordered native state to the less ordered one occurs.

Frequently the process of denaturation is considered as an abrupt cooperative transition (the transition of the type "all or nothing").

This notion is an oversimplification, but one-stage model that could be described by the irreversible reaction of the first order $N \xrightarrow{\kappa} D$, where N and D denote native and denaturated proteins, is a good approximation for many problems. A fraction of the denaturated protein is expressed as

$$F_d = 1 - \exp\left(-\int_0^t \kappa dt\right),$$

where the reaction rate dependence on the temperature is assumed to obey the Arrhenius model:

$$\kappa(T) = A\exp\left(-\frac{E_a}{RT}\right),$$

here E_a is the activation energy.

The application of the kinetic models of higher order that explicitly consider the intermediate states of protein between the native and denaturated ones to the process of the protein denaturation is certainly justified if a distinction is made between the states of the protein after the (reversible) unfolding U and the final denaturated state D that arises due to aggregation (either ordered one called gelation or the random aggregation called coagulation), chemical modification of the amino acid residues, or by other reasons. A two-stage model of Lamri and Eiring $N \Leftrightarrow U \rightarrow D$ should be used in this case. A description of more complex models of the denaturation could be found in the cited paper. The denaturated state of the protein could be nonunique.

Protein denaturation can be the rate-limiting stage in the hyperthermic cell death that is confirmed by the experimentally measured correlation between the cell death and the protein denaturation. Protein denaturation during freezing or thawing can be also caused by proteases leaking from lyposomes that lost membrane integrity.

Membrane Behavior

The cell membrane is a bilayer of amphiphilic (having both hydrophobic and hydrophilic parts – from Greek *amphi*, on both ends, and *phillos*, loving) molecules – lipids that include phospholipids, glycolipids, and steroids; the membrane could contain up to a hundred of different lipid molecules and a number of proteins – both peripheric and integral. For example, in human erythrocytes – red blood cells (RBC) – membrane lipids (mostly unestherified cholesterol and phospholipids, with small amounts of free fatty acids and glycolipids) compromise about 40% of its mass while membrane proteins about 52% of mass.

The membrane lipids have a small polar head and a long nonpolar (hydrocarbon) tail. Most lipids are nearly insoluble in water (solubility values are between 10^{-10} and 10^{-6} moles per litre) and tend to form aggregates spontaneously. Bilayers consist of two opposing hydrocarbon mono- layers, separated from the surrounding water by two layers of the polar lipid headgroups.

The membrane forms the boundary between the intracellular and extracellular compartments and mediates transport into and out of cell. The membrane also provides the mechanical coupling between the cell and its environment - the neighbor cells and the extracellular matrix. The cell membrane is a composite structure described as 2D "*fluid mosaic*" with inclusion of various membrane-associated proteins. Among the membrane proteins are ion channels, receptors, and signaling macromolecules that are diffusing in the lipid bilayer or linked to the cytoskeleton or to the extracellular matrix. In erythrocytes, for instance, peripheral proteins - spectrin, ankyrin, actin, adducin - found on the cytoplasmic side of the membrane form a mesh-like network of microfilaments that provide the mechanical strength of the membrane, in particular, cellular deformability.

The interaction of the molecules that constitute the membrane with the water molecules determines both the membrane structure and stability; some water molecules - the hydration shells around the polar parts of the lipid molecules and proteins as well as the singular water molecules in the hydrocarbon zone of the membrane - are osmotically inactive. The surface of the membrane is usually charged (in the average) negatively. Depending on the water content in the system and the electrolyte concentration, the different mesophase structures could be observed; the biological membrane in the normal state is a lyotropic liquid crystal.

The classic fluid mosaic model of Singer and Nicolson assumes that only one type of lipid molecule is present in the membrane; its generalization to the case of two different lipids extends significantly to the class of problems that could be considered rigorously, in particularity, it becomes possible to correctly describe a deformation of the membrane and the phase separation. The heterogeneity and the dynamic character of the membrane are essential for its function.

In case of the membrane consisting of incompatible amphiphiles A and B, a phenomenological parameter ϕ is introduced, defined as the fraction of the surface area occupied by the component A, that is, ϕ = *area(A)/(area(A) + area (B))*. For the phase-separated membranes under weak tension, the total free energy is written as the sum of the contributions from the bending elasticity F_b, the membrane surface tension F_λ, and the line tension between phases F_σ: $F = F_b + F_\lambda + F_\sigma$, and these contributions are defined as

$$F_b = \sum_{i=A,B} \iint_{S_i} \frac{\kappa_i^c}{2}(2H - c_0)^2 dS, \quad F_\lambda = \sum_{i=A,B} \lambda \iint_{S_i} dS, \quad F_\sigma = \sigma \int_{\partial S} dl,$$

where i = *A, B* denotes the domains of the membrane occupied by the corresponding components, κ_i^c is the bending rigidity, λ and σ are the surface and line tension, H is the mean membrane curvature, and c_0 is the spontaneous curvature.

Ni et al. considered the case of surface tension λ, spontaneous curvature c_0, and Gaussian curvature rigidity being the same for the both components (the only differences are in the bending rigidity κ_i^c) and found several stable patterns of phase arrangements, including a linear array of A-phase stripes and a hexagonal array of circular A-phase caplets in the matrix of B-phase, with shape of domains being strongly influenced by the composition ϕ. The authors also found that the different bending rigidities of the lipid components could induce the first-order transition between the phases. The case of different signs of spontaneous curvature of the components of the two-phase membrane was considered by Harden et al.

Sunil Kumar et al. studied modulated phases in the multicomponent fluid membranes; two-component (A, B) bilayer, and three-component (A, B, C) monolayer membranes were considered. For the former, the authors found that two square and three hexagonal phase separation pattern are possible, for the latter in addition two stripe phases and homogeneous A-rich, B-rich, and C-rich disordered phases were observed. In certain restricted regions of parameter space, the behavior of the three-component monolayer is similar to the behavior of the two-component bilayer. The authors note that the two phase coexistence between gel and fluid phases is more difficult for the analysis in comparison with the fluid–fluid coexistence, since the shear elastic properties of the gel domains should be taken into account.

The membrane reacts to the temperature variation (*thermotropic mesomorphism* – the order of the membrane constituents changes with temperature): heating results in the formation of the hexagonal structures in the lipid bilayer, cooling leads to the transition of the membrane lipids into the gel phase ("the liquidsolid transition," also called the *main phase transition*). Low temperatures also induce alterations in the lipid composition of the membrane (the degree of nonsaturation). Phase separation is a result of the difference in the phase transition temperatures of different lipid molecules present in the membrane; often it leads to formation of gel-phase rich domains ("*aparticulate domains*"), in which integral proteins have been excluded from the lipid layer; domains have also been detected on whole cells.

The temperature of the main phase transition depends on a number of factors, including the composition of the membrane, the degree of hydration, the presence of impurities in the lipid bilayer, and the composition of the suspending solution.

The transition of the membrane into the gel phase is accompanied by the change of the sizes of the elements of its structure, an appearance of defects between the membrane proteins and the lipid bilayer, increasing of the membrane permeability for ions, and possible leakage of cytoplasmic solutes. The critical temperature of the transition depends on the water content in the system and attains the minimal value when it exceeds the water content that could be bond by the lipid structures as well as on the membrane composition (a fraction of the saturated and unsaturated phospholipids; a well-known example of the composition effect is the variation of the fraction of the unsaturated phospholipids in the cells of the north deer *Rangifer tarandus* leg from the hooves upward in accordance with the variation – by tens of degrees Centigrade – of the characteristic temperature).

Cell dehydration influence the membrane behavior: it increases the main phase transition temperature of lipids. As a result, the phase separation could occur at higher temperatures. Experiments on freezing of prostate tumor cells showed that the onset of the liquid crystalline to gel transition consides with the ice nucleation temperature. It is unclear whether this relation is valid for other types of cells.

Freeze-induced dehydration was shown to result in the alteration of the permeability parameters of the protoplast membrane – protoplast became osmotically unresponsive. Electron microscopy showed that membrane defects such as "*fracture-jump lesion*" related to the existence of aparticulate domains are formed.

The cooling reduces the potential of the muscle membranes and the excitability of the muscle fibers that is explained by the loss of the integrity of membranes and changes in their composition, mainly removing of the hydrophilic lipid components. The membrane fragmentation could be preceeded by the loss of the membrane proteins. Reduction of the mechanical strength of the membranes is confirmed by the results of the centrifugation of the cooled cells.

Chakrabarty et al. studied the variation of the membrane composition during the cryopreservation and found a significant decrease in the content of neutral lipids, glycolipids, and phospholipids. The most significant drop of the concentration among neutral lipids was recorded for steryl esters and for 1-*O*-alkyl-2,3-diacyl glycerol, and for phosphatidyl choline and phosphatidyl ethanolamine among phospholipids. As a result the cholesterol:phospholipid ratio increases as well as the amount of hydrocarbon. In another words, increase of the hydrophobicity of the membrane is observed to be caused by the shedding-off of preferentially specific hydrophilic constituents of the cell membrane. The authors note that such change of the membrane hydrophobicity aiming to counteract the cold stress could not be totally reversible upon thawing.

Temperature lowering also results in the cytoskeleton depolymerization and reduces the ability of the integral proteins to control the intracellular electrolyte concentration that could finally cause the protein denaturation.

Cells in Aqueous Solutions

The dependence of behavior of an isolated cell in the aqueous solution during freezing on the cooling rate could be explained using a simple "*saltwater sack*" model. The process starts with the crystallization of the supercooled solution, resulting in the growth of its concentration since the solutes including ions are repelled from the ice crystallization front.

The nucleation rate initially increases with the degree of supercooling, reaching a maximal value and thereafter decreases due to the drop of the mobility of the water molecules. In growth of the polycrystals, the initial and secondary nucleations are distinguished; under the latter the processes at the surface of the existing grains are meant.

The values of supercooling measured for different samples are rather high: for example, up to –10 °C for the rabbit corneal tissue, up to –20 °C (in the absence of the extracellular ice) for the human erythrocytes, up to –7 °C and –18 °C for the vesicles of lemon in the presence and absence of the extracellular ice, respectively.

The morphology of the growing crystal (including, in the case of the growth of dendrites if any, a fractal dimension) is determined, first of all, by the degree of supercooling. The extensive study using polarized light of the ice crystallization and growth in pure water in the wide range of the supercooling from 0.1 °C to 30 °C for the case of the heterogeneous nucleation proved the existence of at least six distinct crystal shapes, depending primarily on the supercooling value:

1. The dense branched structure;
2. The dendrite;
3. The morphologically stable needle;
4. The fractal needle-like branch;
5. The compact needle-like branch;
6. The plate

and of the different growth regimes: convection–diffusion, diffusion, kinetic. The crossover from the diffusion-limited to the kinetics-limited regime also was investigated. The fractal dimension of dendrites d_f turned out also to depend on the initial supercooling of water: in the temperature range [0.3– 4K], d_f decreases monotonically from 1.6 down to 1 that corresponds to the continuous "pulling" of the dendrite with developed side branches into the smooth needle-like crystal.

Cell Dehydration

When a cell is cooled slowly, its destruction is caused by the cell dehydration (*exosmosis* - the water is leaving the cell both by the agency of diffusion across the bilipid cell membrane and by the hydrodynamic flow through the pores in the membrane), which results in the growth of the intracellular concentration of solutes and the shrinking of the cell. Diffusion of water molecules across the bilipid layer consists of three steps: (1) partitioning of the water molecule from the aqueous phase into the lipid phase; (2) diffusion through the lipid layer; (3) return from the lipid to the aqueous phase - all of which could be associated with some activation energy.

Considering the intracellular solution, it is necessary to distinguish the mean salt concentration

$$\bar{c}^{i}_{\text{salt}} = \frac{\text{moles of salt}}{\text{total cell volume}} = \frac{N_{\text{salt}}}{V}$$

and the actual concentration c^{i}_{salt}

$$c^{i}_{\text{salt}} = \frac{\text{moles of salt}}{\text{cytosol or osmotically active volume}} = \frac{N_{\text{salt}}}{V - V_b},$$

where V_b is the osmotically inactive volume. Introducing the fraction of the osmotically inactive cell volume $\varphi = V_b/V$, the relation between these concentrations could be written as

$$c_{\text{salt}}^{-i} = c_{\text{salt}}^{i}(1-\phi).$$

If a semipermeable solute is present in the solution, its mean cellular concentration will be $c_s^{-i} = N_s/V$. To determine the actual solute concentration, it is necessary to use the Henry law of absorption with coefficient k:

$$c_s^i(V - V_b) + c_s^i \kappa s_m + c_s^i K V_b = N_s$$

or

$$c_s^{-i} = c_s^i(1 - \phi + \kappa\alpha + K\phi),$$

where $\alpha = s_m/V$ is the specific surface area of organelle membranes and K is the partition coefficient into organelles.

For a continuous volumetric density field (e.g., mass density or the solute concentration) $\psi(r, t)$, its total amount within the moving control volume $V(t)$ is defined as

$$\Psi(t) \equiv \int_{V(t)} \psi(r,t)dV$$

and its variation in time, accounting for the fact that cellular membrane could be considered as non-material interface as

$$\frac{d\Psi}{dt} = \int_{V(t)} \left(\frac{\partial\psi}{\partial t} + \mathrm{v}\cdot\nabla\psi\right) dV + \int_{\partial V} dS\cdot(u - \mathrm{v})\psi,$$

where ∂V is the volume boundary moving with the velocity $\boldsymbol{u}$. For the spherical cell these relations are simplified to

$$\Psi(t) \equiv \int_0^{R(t)} \psi(r,t)r^2 dr$$

and

$$\frac{d\Psi}{dt} = 4\pi \int_0^{R(t)} \frac{\partial\psi}{\partial t} r^2 dr + 4\pi R(t)^2 u(t)\psi(R(t),t)$$

When the salt or other solute concentration is chosen as $\boldsymbol{\varnothing}$, the above relations give the rate of change of the corresponding substance in the cell:

$$\frac{dN}{dt} = 4\pi R^2 \left(\bar{D}\frac{\partial\bar{c}}{dr} + uc\right) \Big| r = R.$$

The quantity in the parentheses in the above expression is the flux (per unit area) of the substance, which is zero for the salt while in case of the semipermeable solute it is related to the its distribution coefficient at the membrane – the measure of its ability for permeation.

Assuming the intracellular solution to be ideal, that is, the difference in the chemical potentials of the cytoplasm and the water to be proportional to ln(1 − χ), where χ is the molar concentration of solutes, a depression of the crystallization temperature $\Delta T \propto \chi$ is expressed as

$$\Delta T = \frac{RT_f}{L}\chi = -K_{\chi},$$

where L is the latent heat of crystallization, T_f is the freezing temperature of the pure liquid, and K is the so-called cryoscopic constant.

This effect is the colligative one, that is, determined by the *quantity* of solutes and does not depend on their nature. Usually the solutes are divided into three groups:

1. Salts – small solutes, charged
2. Sugars and other uncharged medium-sized molecules
3. Macromolecules

An additional effect on the freezing temperature observed in the case of macromolecules is related to the properties of their surface. The effect of the solutes on the structure of the hydrogen bonds net and the crystallization temperature has been studied experimentally by M.B. Baker and M. Baker for the small water drops and found in many respects to be similar to the effect of the higher pressure; it is even possible to express the solute content in terms of an approximate osmotic pressure that is proportional to the logarithm of the water activity in the solution. The effect of the lipid membrane on the freezing point depression depends on the hydration state of the former.

There is a linear relationship between the equilibrium melting point and the homogeneous nucleation temperature of various aqueous solutions $\Delta T_{f,hom} = \lambda\Delta T_m$, where $\Delta T_{f,hom} = T_f^0 - T_f$ and $\Delta T_m = T_m^0 - T_m$, T_f^0 and T_m^0 are the homogeneous nucleation temperature and the equilibrium melting point of pure water, respectively, T_f and T_m are the homogeneous nucleation temperature and the equilibrium melting point of the solution. λ could be considered as the measure of the solute-specific supercooling capacity. Kimizuka et al. measured the value of λ for the solutions of small molecules such as salts, alcohols, sugars and found a linear relationship between λ and the logarithmic value of the solute self-diffusion coefficient D_0. In a recent paper, Kimizuka et al. suggested an explanation of the observed correlation, involving the expression for the nucleation rate J:

$$J = \frac{k\Gamma_z}{\eta(T)} \exp\left(\frac{-\Delta G^{nucl}}{kT}\right),$$

where ΔG^{nucl} is the activation energy for nucleation, Γ_z is the Zeldovich constant, and η is the solution viscosity. The homogeneous nucleation temperature could be found from the condition that the number of nuclei per unit volume becomes equal to 1 and thus evidently depends on the nucleation rate. Assuming further that the Stokes–Einstein relation is valid in the supercooled fluid, the authors get the needed relationship between λ and D_0 through the temperature-dependent viscosity of the solution $\eta(T)$.

The value of λ for polymer solution is rather high and increases with the molecular weight, indicating that these solutions allow deep supercooling.

The concentration of the extracellular solution increases due to the solute rejection from the ice crystallization front – the so called constitutional supercooling arises – until the eutectic temperature is reached at which the remaining solvent and solute crystallize together. In the case of sodium chloride this temperature is –21.1 °C and the eutectic mixture contains 31% NaCl. Additional depression of the freezing point for cytoplasm is related to the interaction of the ions in the solution with the charged sites of the proteins present inside the cell and bringing the proteins into the solution. The reverse process – proteins salted back out of the solution – is observed upon thawing when the intracellular solution, due to the ice melting, is diluted back to isotonic.

Mazur et al. reported in a series of papers the results of their search for correlations between the temperature at which the *intracellular ice formation* (IIF) occurs ("*flashing temperature*") and the state of the external solution. The authors studied the IIF in the mouse oocytes and in stage I and II *Xenopus laevis* oocytes in the solutions of glycerol or ethylene glycol. They found that the flashing temperatures in *Xenopus* are significantly higher than those observed in mouse oocytes, which is attributed to the much larger size of the former. The IIF in mouse oocytes is found to occur in the narrow range of unfrozen fractions of the external solution that allowed the authors to state that these entities are casually related and to introduce the notion of the critical unfrozen fraction. The results for *Xenopus* in the EG solutions do not show such clear trend, while the IIF in this case could be correlated with the narrow range of the solution concentration (solution osmolality).

Mazur and Kleinhans suppose that in the case of *Xenopus* oocytes in EG solution, the dominant damage mechanism is the chemical damage of the membrane in the highly concentrated solutions.

The authors also stress that the correct interpretation of the experimental data depends on the accuracy of the phase diagrams used to determine the solution properties. In particularity, they found that the phase diagrams for the ternary solutions EG/NaCl/water obtained by the synthesis of the data for binary solutions EG/water and NaCl/water are more accurate (i.e., provide more consistent interpretation of the measured data) than those obtained experimentally directly for the ternary solutions.

The specific mechanism that could be responsible for the correlation of the unfrozen fraction of the external solution with the IIF (as well as for the existence of the critical value of the unfrozen fraction) remains obscure. It is assumed that this mechanism could involve the contact of the external ice with the existing pores in the membrane and the membrane defects or severe deformation of the membrane surface attempting to conform its shape to the available small extracellular space that is still in the liquid phase.

Preventing the ice formation inside the cell is by all means favorable for the cell survival; however, the growth of the solute concentration results in "solution effect injury": the proteins denaturation, the lysis (a breaking of the cellular membrane), injuries to the mitochondria and the nucleus. Some biological objects are relatively resistant to solute effects, for example, eight-cell embryos. The membrane behavior is sensitive to the temperature of the solution crystallization.

Two-compartment model of the cell's dehydration and shrinking (to be more exact, the cell volume change) in the freezing solution suggested by Mazur over 40 years ago still forms the basis of many modern models. Mazur made a number of assumptions:

1. The plasma membrane remains intact during cooling
2. The membrane is permeable for water only
3. Protoplasm is an ideal solution, so Raoult law is applicable
4. The extracellular solution is in the thermodynamic equilibrium with ice
5. The intracellular temperature is equal to that of the environment
6. Substitution of the molar volume of water for the partial molar volume does not result in large errors
7. The cooling rate is constant

Mazur also used Arrhenius-type temperature dependence of the hydraulic permeability presented below and derived equations for the cell volume evolution as a function of time or of temperature.

Extension of the model to the case of the solution containing an intracellular (permeating) cryoprotective agent adds two more parameters: the membrane permeability for the cryoprotectant and a parameter describing the solute interaction with the water molecules. Generalization of the model to nonideal nondilute solutions was performed by Studholm. Thus, for the ternary solution consisting of a permeable solute (CPA), a nonpermeable solute (NaCl), and water, the total transmembrane volume flux J_V (thus also the rate of the cell volume change) and the permeable solute flux J_s are written as follows:

$$J_V = \frac{1}{A}\frac{dV}{dt} = L_p[(p^e - p^i) - R_g T(M_n^e - M_n^i) - \sigma R_g T(M_s^e - M_s^i)],$$

$$J_s = A[(1-\sigma)\bar{m}_s J_V + P_s(a_s^e - a_s^i)],$$

where superscripts *e* and *i* refer to the extracellular and intracellular solutions, respectively, *V* is the cell volume, *p* is the hydrostatic pressure, P_s is the permeability coefficient of CPA, σ is the reflection coefficient, *M* is the osmolality, *a* is the activity and

$$\bar{m}_s = \frac{m_s^e - m_s^e}{\ln m_s^e - \ln m_s^e},$$

m is the molar concentration.

The Mazur model exploits the Kedem–Kachalsky equations for the non-electrolyte transport through the permeable membrane that divides two compartments containing the well-mixed ideal dilute solutions:

$$J_v = L_p \Delta P - L_p \sigma \Delta \Pi \quad \text{...(1)}$$

$$j_s = \omega \Delta \Pi + (1-\sigma)\bar{c} J_v, \quad \text{...(2)}$$

where J_v is the volume flux, j_s is the solute flux, L_p is the hydraulic permeability (filtration coefficient), σ and ω are the reflection and transmission coefficients, ΔP and $\Delta \Pi$ are the drops of the hydraulic and osmotic pressures over the membrane, and c^- is the average concentration.

Equations (1,2) were derived in the frame of the linear thermodynamics theory of irreversible processes. "Mechanistic" derivation of the equations describing the transport through the membrane proposed by Kargol and Kargol provides a simple interpretation of the coefficients entering the equations; the membranes having pores of different radii were also considered in the cited papers.

The osmotic transport of water out of the cell (exmosis) is limited by the membrane permeability that depends on the temperature:

$$L_p = L_{p0} \exp\left(-\frac{E_a}{R}\left(\frac{1}{T_0} - \frac{1}{T} \right) \right) \qquad ...(3)$$

and serves as a basic parameter of the Mazur's model, E_a is the activation energy.

The water transport parameters of the cell membrane - the hydraulic permeability and its activation energy - are shown to alter due to the cold shock and, to a lesser extent, due to osmotic shock.

Probably, the most general up-to-day formulation of the problem for the cell in the multicomponent solution given by Ateshian et al. takes into account the deviation of the segregation coefficients of solutes from unity and the mechanical stress in the membrane.

The authors considered a general case of the transport of uncharged solutes in the uncharged porous media using the mixture theory. The mixture contains a solid matrix (α = s), solvent (α = w), and solutes ($\alpha \neq$ s,w). The momentum equations for the mixture as a whole and for the movable constituents (solvent and solutes) are written as

$$-\nabla p + \nabla \cdot \sigma = 0, \qquad ...(4)$$

$$-\rho^{\alpha} \nabla \tilde{\mu}^{\alpha} + \sum_{\beta} f_{\alpha\beta}(\mathrm{v}^{\beta} - \mathrm{v}^{\alpha}) = 0, \qquad ...(5)$$

where p is the fluid pressure, σ is the effective stress in the solid matrix, $\tilde{\mu}^{\alpha}$ is the chemical potential of the constituent α, v^{α} is the velocity of the constituent α, $f_{\alpha\beta}$ is the diffusive drug coefficient between the components α and β (evidently, $f_{\alpha\beta} = f_{\beta\alpha}$).

The mixture is assumed to be dilute and ideal, and so the solvent and solute chemical potentials are as follows:

$$\tilde{\mu}^{w} = \tilde{\mu}_0^{w}(T) + \frac{1}{\rho_0^{w}}\left(p - R_g T \sum_{\beta \neq s,w} c^{\beta} \right), \qquad ...(6)$$

$$\tilde{\mu}^{\alpha} = \tilde{\mu}_0^{\alpha} \frac{R_g T}{M_{\alpha}} \ln \frac{c^{\alpha}}{\kappa^{\alpha} c_0^{\alpha}}, \qquad ...(7)$$

where c^{α} is the solute concentration, R_g is the universal gas constant, M_{α} is the molecular weight of the solute α, κ^{α} is the partition coefficient accounting for the steric volume exclusion effects (molecules of solute may have no access to small pores), ρ_0^{w} is the true density of the solvent.

Ateshian et al. simplified the general equations for the momentum conservation, neglecting the solute-against-solute drag and accounting for the smallness of the volume fraction of the solutes in comparison to that of the solvent, $\phi^{\alpha} \ll \phi^{w}$. The latter in the general case of the deformable porous media depends on the solid matrix dilatation as

$$\phi^{w} = 1 - \frac{1-\phi_0^{w}}{\det F},$$

where ϕ_0^{w} is the porosity at zero deformation and F is the deformation gradient of the solid matrix.

The momentum equations are supplemented by the continuity (mass conservation) equations for the mixture as a whole and for the individual components:

$$\nabla \cdot \left[\mathrm{v}^{s} + \phi^{w}(\mathrm{v}^{w} - \mathrm{v}^{s}) + \sum_{\alpha \neq s,w} \phi^{\alpha}(\mathrm{v}^{\alpha} - \mathrm{v}^{s}) \right] \approx \nabla \cdot [\mathrm{v}^{s} + \varphi^{w}(\mathrm{v}^{w} - \mathrm{v}^{s})] = 0,$$

$$\frac{\partial \phi^{w} c^{\alpha}}{\partial t} - \nabla \cdot \left[\phi^{w} D^{\alpha} \nabla c^{\alpha} - \phi^{w} c^{\alpha} \left(\mathrm{v}^{s} + \frac{D^{\alpha}}{D_0^{\alpha}} (\mathrm{v}^{w} - \mathrm{v}^{s}) \right) \right] = 0. \quad \ldots(9)$$

To close the system of equations, the authors formulate the conditions at the interface that is defined somewhere on the solid matrix.

The general approach is applied to the biological membrane that is assumed to be thin and to have the spherical shape. The former means that all gradients are reduce to the scalar differences of the variables across the membrane. The smallness of the thickness and sphericity allow to write

$$\Delta p - \Delta \sigma_{rr} - \frac{T'}{r_0} = 0,$$

where r_0 is the cell radius, T' is the surface tension, and σ_{rr} is the stress in the radial direction. It was assumed that the mechanical reaction of the membrane has a threshold nature (the surface tension is considered to become zero when the membrane surface area falls down below the specified critical value).

Using conditions at the interfaces between the cytoplasm and the membrane and between the membrane and the extracellular solution, the authors get the general equations governing the cell volume evolution and the intracellular concentration of the solutes, which for the case of one permeating ($\alpha = p$) and one nonpermeating ($\alpha = n$) solutes reduce to the following equations:

$$\frac{dV}{dt} = -AL_p\left(\frac{2T'}{r_0} - R_gT\left[c_i^p - c_e^p + c_i^n - c_e^n - (1-\sigma^p)\left(\frac{c_i^p}{\kappa_i^p} - c_e^p\right)\right]\right), 10$$

$$\frac{dn_i^p}{dt} = -AP^p\left(\frac{c_i^p}{\kappa_i^p} - c_e^p\right) + (1-\sigma^p)\frac{1}{2}\left(\frac{c_i^p}{\kappa_i^p} - c_e^p\right)\frac{dV}{dt}. \quad ...(11)$$

Here the subscript *i* and *e* refer to the intracellular and extracellular solution, respectively, *A* is the cell surface area, P^p is the membrane permeability for the solute *p*, and the hydraulic permeability of the membrane is expressed as

$$L_p = \frac{1}{\frac{1}{L_{p0}} + R_gT\frac{(\kappa_i^p - 1 + \sigma^p)(1-\sigma^p)}{2P^p}\left(\frac{c_i^p}{\kappa_i^p} + c_e^p\right)}. \quad ...(12)$$

The expressions for the equilibrium values of the cell volume and of the intracellular solute concentration could be found in the cited paper. As expected, these values do not depend on the transport parameters L_p, P^p, σ^p.

These equations reduce to the three-parameter (L_p, P^p, σ^p) Kedem–Katchalsky model if

1. The membrane surface tension is neglected $T' = 0$
2. The permeating solute partition coefficient is unity within the cytoplasm $\kappa_i^p = 1$.

However, in the reduced form of the equations based on the general mixture theory, the hydraulic permeability L_p, in contrast to the phenomenological model, depends on the solute concentrations according to (12).

Nonspherical Cells

Batycky et al. considered cells of arbitrary shape. Note that the initially spherical cell could change its shape under the nonuniform osmotic stresses. In the spherical cell due to the incompressibility of the cytosol fluid, its velocity (the radial component, thanks to the symmetry) should satisfy the continuity equation

$$\nabla \cdot \mathrm{v} = \frac{1}{r^2}\frac{d}{dr}\left(r^2 v_r\right) = 0,$$

which has a general solution

$$v_r(r) = \frac{A}{r^2},$$

where *A* should be zero to avoid the singularity at the origin, thus

$$\mathrm{v} = 0.$$

Therefore, the cytosol does not move due to the expansion or contraction of the membrane.

In contrast to the case of the spherical case, the hydrodynamic velocity of cytoplasm is not zero; still, however, its scale is governed by the membrane motion, the quasi stationary analysis remains valid; the solution of the equations in the general case requires the numerical approach. The authors report that they have derived approximate (analytical) results for the practically important case of the cylindrical cell.

Membrane-limited vs. Diffusion-limited Transport

The described zero-dimensional model neglects the diffusion processes in the cell cytoplasm assuming the cytoplasm to be a well-mixed solution. This assumption is valid, for example, for leucocytes and lymphocytes, but is not reasonable for erythrocytes. A one-dimensional model that incorporate intracellular diffusion developed by Carnevale allows to distinguish the dehydration regimes limited by the membrane transport and by the diffusion processes in the cell.

Diffusion inside the cell is govern by the usual parabolic equation, which for the spherical cell is expressed as

$$\frac{\partial c_w(r,t)}{\partial t} = \frac{1}{r^2}\frac{\partial}{\partial r}\left(D(r,t)r^2\frac{\partial c_w(r,t)}{\partial r}\right). \quad ...(13)$$

The water flux J_w through the cell membrane should be provided by the diffusive flux from the cell's interior. These fluxes, however, differ due to the cell membrane movements. Continuity requirements yield the following boundary condition for the concentration at the membrane surface $r = R$:

$$-D\frac{\partial c_w(r,t)}{\partial r}\bigg|_{r=R} = J_w + c_w\frac{dR}{dt}. \quad ...(14)$$

Zero diffusive flux is specified at the cell's centre

$$\frac{\partial c_w(r,t)}{\partial r}\bigg|_{r=0} = 0. \quad ...(15)$$

The equation for the membrane velocity follows from the incompressibility of the intracellular fluid

$$\frac{dR}{dt} = -J_w v_w, \quad ...(16)$$

where v_w is molar specific volume of water, $J_w = (L_p/v_w^2)\Delta\mu$, $\Delta\mu$ is the difference of the chemical potentials across the membrane $\Delta\mu = \mu_{ice} - \mu_w$.

The incompressibility of the cytosol and conditions under which only water transport across the membrane occurs lead to the on-to-one correspondence between the cell volume and the amount of water contained in the cell

$$c_w = \frac{(V_0 - V_b)c_{w0}v_w - (V_0 - V)}{(V - V_b)v_w},$$

where c_{w0} and V_0 are the initial values of the cell water concentration and the cell volume, respectively, and V_b is the osmotically inactive volume.

In computations, Carnevale set the diffusion coefficient to be constant and estimated its value according to the Stokes–Einstein relation (4). The apparent hydrodynamic radius of the water molecule equals to $a = 1.4 \times 10^{-10}$ m; for temperature dependence of the water viscosity, Vogel–Fulcher–Tammann-type dependence (6) was used and the viscosity of the suspension was considered as that of a suspension of rigid spheres and estimated as

$$\eta = \eta_w \exp\left(\frac{\kappa_E \phi_{\text{sph}}}{1 - \lambda\phi_{\text{sph}}}\right),$$

where $\kappa_E = 2.5$ is the Einstein shape factor for spheres, $\lambda = 0.609375$ is the hydrodynamic interaction parameter, and ϕ_{sph} is the volume fraction of spheres. The latter was calculated under assumption that the diffusing rigid sphere is the hydrated salt ion, thus

$$\phi_{\text{sph}} = c_s(v_s + hv_w),$$

where c_s is the salt concentration, v_s is the molar specific volume of salt, and h is the effective number of the water molecules in the hydration shell.

The chemical potential of the extracellular water is determined by the Clausius–Clapeyron relationship as

$$\mu_{\text{ice}} = \mu_0 + L\left(\frac{T}{T_0} - 1\right),$$

where T_0 is the equilibrium freezing temperature of water and μ_0 is the chemical potential of water at T_0.

Under the usual assumption of the validity of the Raoult law (i.e., considering cytosol as an ideal solution), one can write for the chemical potential of the intracellular water

$$\mu_w = \mu_0 + R_g T \ln(\chi_w),$$

where the mole fraction of the intracellular water

$$\chi_w = \frac{c_w}{c_w + \sigma_s c_s},$$

where $\sigma_s = 2$ is the salt dissociation constant.

A few characteristic times could be defined for the problem of the cell dehydration. The first one characterizes the duration of the whole process

$$\delta t_B = T_0,$$

where $B = -dT/dt$ is the externally specified cooling rate.

The characteristic time of the intracellular diffusion is

$$\delta t_D = \frac{R^2}{D_0},$$

where D_0 is the characteristic diffusion coefficient (e.g., at the cell centre), while that of the membrane transport is

$$\delta t_M = \frac{R v_w}{L_p R_g T}.$$

Since the cell dehydration involves both the membrane transport and intracell diffusion, water loss will be insignificant and the probability of the IIF high if either of the two characteristic times δt_D or δt_M will be large: $\delta t_D \gg 1$ or $\delta t_M \gg 1$.

Intracellular Ice

When the cooling is fast, the water has not enough time to leave the cell and formation of the intracellular ice causes the cell destruction mechanically disrupting the cell membrane and the cytoskeleton. IIF observed in experiments is often referred to as "*flashing*" or "*darkening*" of cells. The lethal character of IIF is known for a long time.

The temperature at which IIF occurs and amount of water within the cell at that moment determine the total amount of intracellular ice. Conversion of more than 5–10% of normal intracellular water into ice is considered lethal to most cells while lesser amounts of ice are often tolerated. The effect of the nucleation temperature on the thermotropic behavior of the membrane is usually attributed to differences in the extent of the cell dehydration.

The probability of the spontaneous nucleation depends on the cytoplasm volume, the degree of supercooling, the solute concentration, and the fluid viscosity. Recently, as was already mentioned, Mazur et al. have studied the relation of the IIF to the fraction of the unfrozen water.

A thermodynamic model of IIF accounting for the heterogeneous nucleation at the cellular membrane and at the particles contained in the cytoplasm was developed by Toner et al. The authors starting from the classic nucleation theory as a sequence of bimolecular reactions $\alpha + \beta_m \Leftrightarrow \beta_{m+1}$ found the rate of homogeneous nucleation I^s as a function of two parameters – thermodynamic Ω and kinetic κ ones:

$$I^s = \Omega \exp(-\kappa/(\Delta T^2 T^3)),$$

where

$$\Omega = 2(\upsilon^{\beta})^{1/3}\frac{(\sigma^{\alpha\beta}\kappa T)^{1/2}}{(\eta\upsilon^{\alpha})N_{\upsilon}}, \qquad \kappa = \frac{16\pi(\sigma^{\alpha\beta})^3 T_f^4}{(3\kappa L^2)}, \qquad ...(17)$$

ΔT is the supercooling, υ^{α} is the molecular volume of the phase α, $\sigma^{\alpha\beta}$ is the free energy of the ice–water interface, L is the latent heat, η is the cytoplasm viscosity. The latter is approximated as

$$\eta/\eta\hat{w} = \exp\left(\frac{5}{2}\frac{\phi_m}{(1-Q\phi_m)}\right),$$

where $\eta\hat{w}$ is the viscosity of pure water, ϕ_m is the total fraction of solutes (e.g., salts and proteins), and Q is the interaction parameter. To approximate the viscosity dependence on the temperature, either Vogel–Fulcher–Tammann law $\eta \propto \exp(E/(T - T_0))$ or polynomial approximation is used.

The interfacial free energy for ice/water system could be measured by different methods, most of which exploit the Gibbs–Thompson relationship for the depression of the melting temperature due to the curvature of the crystallization front; the recent measurements of the supercooling needed for the freezing in the porous media by Hillig gave the value 31 ± 2.7 mJ m^{-2}.

Toner et al. also proposed a modification of the parameters Ω and κ for the case of the heterogeneous nucleation, which consists in the introduction into (17) factors $(f(\theta_0))^{1/6}$ and $f(\theta_0)$, respectively, $f(\theta) = (1 - \cos\theta)^2 (2 + \cos\theta)/4$, and θ_0 is the contact angle between the ice cluster and the inner membrane surface, which depend linearly on the solution concentration. The proposed model is highly sensitive to the values of κ and η. Later this model was combined with the model of the transport through the membrane and used to study cells behavior in the mixture "water–NaCl–glycerol."

Ice crystals at the end of freezing could be too small to injure the cellular organelles, but the recrystallization during thawing leads to the fatal increase of the crystal size. At low temperatures the passive ion transport is essentially unaltered, but the efficiency of the active ion transport through the membrane, which is based on the metabolic reactions, drops. The cell could be destructed also in the case of the eutectic crystallization.

Koshimoto and Mazur drew attention to a significant (eightfold) discrepancy between the cooling rate predictions of IIF using the model of the kinetic cell dehydration and the results derived from the experimental survival curves vs. the cooling rate for the mouse spermatozoa. The somewhat smaller discrepancy between the cooling rates needed to produce IIF in the mouse spermatozoa and those inferred from the survival curves was reported by Devireddy et al. Curry et al. found that the calculated optimal cooling rates fail to conform with the observation for the ram and human spermatozoa, with a difference as high as four orders of magnitude in the critical cooling rate for the ram spermatozoa.

Mazur and Koshimoto list possible causes for this discrepancy. The first is the range of the reasonable values for the poorly known critical parameters - the hydraulic permeability of the cell membrane L_p and its activation energy E_a - is large enough to account for the observed discrepancy. Spermatozoa, being the extremely permeable cells with very low activation energy, are known to be difficult subjects for the L_p measurement. These parameters are also extremely sensitive to the cryoprotective solution composition: for example, in the mammalian embryos, the hydraulic permeability reduces by a factor of 2 or more in the presence of the intracellular CPA. Gilmore et al. have also reported significant increase of the activation energy for the hydraulic permeability at low temperatures when glycerol, DMSO, ethylen glycol, or propylene glycol are present in the solution. Greater changes are possible at lower temperatures, but it is difficult to measure the membrane parameters below the freezing point. There are two routes for water passage into the cell - by diffusion through the lipid bilayer and through the pores in the membrane. Not surprisingly, the porous membranes in comparison with the pure phospholipid membranes have higher osmotic permeability and lower activation energy. Curry et al. suppose that their results support the hypothesis of Muldrew and McGann on the role water movement in the cell cryoinjury.

Next, the nucleation temperature needed for calculations is usually chosen somewhat arbitrarily. The authors also note that injury could result from the internal freezing in the cell organelles, such as mitochondria rather than IIF in the whole cytoplasm. If so, the parameters of the organelle's membrane rather than those of the cell membrane should be used. Recently, Morris reported the results of the study of IIF in human spermatozoa in the wide range of cooling rates from 0.3 to 3,000 °C min^{-1}. Using cryoscanning electron microscopy (CryoSEM), the author did not find the evidence for the IIF within the resolution limits of the technique. The author asserts that in earlier studies where no direct evidence for the IIF was presented, the drop of the survival curve vs. the cooling rate could be wrongly interpreted as the IIF, while the cell death was caused by a long period of exposure to a hypothonic solution during warming.

No intracellular ice was observed in experiments by Fonseca et al. who studied cryoconservation of *Lactobacillus delibrueckii* subsp. *bulgaricus* - lactic acid bacteria widely used in milk industry - in glycerol solutions using three freezing protocols and two storage temperatures. The authors' aim was to separate the effects of freezing kinetics and the storage temperature. The authors note that because of the high viscosity of the water–glycerol mixture, the growth of ice becomes diffusion-limited and the amount of ice formed becomes dependent on the cooling rate.

The cells' damage occurs during warming, which includes the crystallization of the residual water that was kinetically arrested during cooling. Fonseca et al. found that recrystallization during warming is especially harmful if the samples were stored at temperature above the glass transition temperature.

The role of the solution effect injury was studied in using an imitation of the cell's environment evolution during cryoaction: increasing the solution concentration ("crystallization of the solution") followed by the solution dilution ("thawing"). A substitution of the lithium salt that causes a more strong lyotropic effect instead of NaCl results in the more severe cell damage.

Wolfe and Steponkus in their study of the mechanical properties of the plasma membrane of isolated plant protoplasts also used transfer of the sample to hypertonic solution ("*dehydration*") and return to isotonic solution ("*rehydration*") to get the cell's volume and area excursions similar to those occuring during the freeze/thaw cycle. Over short periods (less than a second) the membrane behaves as an elastic two-dimensional media. A different behavior with the resting tension being independent of area was observed over longer time intervals. Stressing that such mechanical properties are not universal since the membrane of protoplast differs greatly from the membranes of red blood cells and some other cell types, the authors suggested an explanation for the long-time membrane behavior based on the slow exchange of the membrane material with some *reservoir* that provides an "extensive" response of the membrane to the tension variation caused by the osmotic manipulations. As a candidate for the role of such reservoir, folds of the membrane or vesicles are considered.

The cell's counteraction to the volume and to the cytoplasm composition changes at the molecular level manifests itself in the activation of the ion channels and the transport mechanisms or in the regulation of the protein expression.

There is an optimal value of the cooling rate for the cell survival. The dependence of the vital cells after the freezing vs. the cooling rate is frequently called the inverse U-shaped curve or "cell survival signature". This appearance of the survival signature is typical for most types of cells; however, there are also exceptions when either the low-cooling-rate or the high-cooling-rate branch of the curve is not distinct, such as yeast cells *Saccharomyces cervisiae* and *Candida utilis* or human leukemia K562 cell.

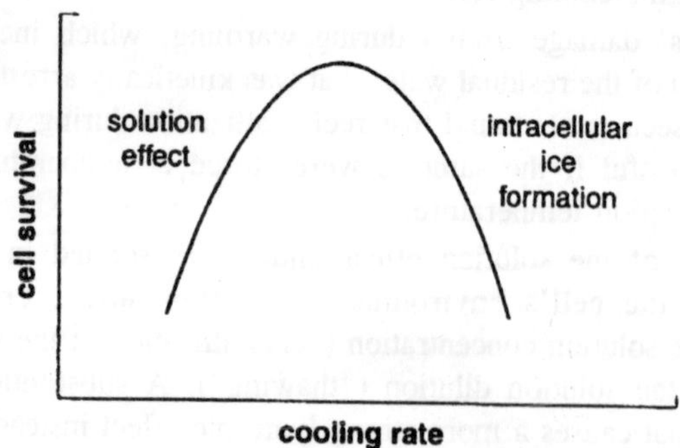

Fig. 4.1. Cell survival signature.

The optimal for the survival value is determined by the characteristic time of the removal of the excess heat due to the crystallization τ_T and the characteristic time of the water diffusion through the membrane τ_D; thus, the two most important cell parameters are its size (or the ratio of the volume to the surface for the cells whose shape is far from the spherical one) and the hydraulic permeability of the membrane. If two cells A and B have similar shape and dimension, but the hydraulic permeability of the first is smaller $L_p(A) < L_p(B)$, the optimal cooling rate for the cell survival will be smaller for cell A.

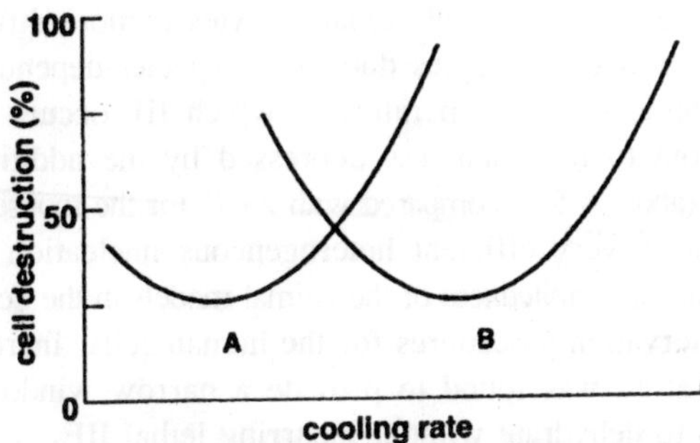

Fig. 4.2. Cell survival signatures for the cells with different hydraulic permeability.

For the erythrocytes having disk-like shape and high permeability, the optimal cooling rate is much higher (up to 10^3 times) than that for other types of cells: for example, the optimal cooling rate for the human erythrocytes is circa 3,000 °C min^{-1}, while it is 1 °C min^{-1} for the bull spermatozoa and 0.5°C min^{-1} for the mouse embryo.

Different stages of embryonic development have different permeability coefficients: the earlier the stage of development, the less permeable are the embryos (morulae are more permeable than eight-cell embryos, eight-cell embryos are more permeable than two-cell embryos, etc.).

One of the most difficult types of cells for the cryopreservation is a mammalian unfertilized egg (oocyte). Mammalian oocytes are characterized by large volume, large amount of cytoplasmic lipids, and presence of MII spindle. Oocytes are highly sensitive to cooling and can be injured even at room temperature due to the delicacy of the miotic spindle apparatus (oocytes are locked in Metaphase II of the cell division cycle) and to the high lipid content of the cells. MII oocytes are more resistant to freezing injury than GV-stage oocytes, which is attributed to the differences in the properties of cytoskeleton elements, namely in the configuration of microtubules and micro-filaments. These elements are straight and rigid in GV-stage while they are flexible in MII-stage oocytes. Thus, hardening of lipids in the cell membrane during freezing more readily results in the permanent damage of the rigid CV-oocyte cytoskeleton.

The uniqueness of the oocytes is that the DNA is held suspended in the cytoplasm on the meiotic spindle and not within the protective confines of the nuclear membrane. Chilling injury may also manifest itself as the thermotropic phase transition in the membrane of the oocyte. Oocytes are also very sensitive to osmotic swelling.

The cryoconservation of the human oocytes is more difficult than that of other mammalian oocytes due to the species-dependent effect of the cryoprotectors: the temperature at which IIF occurs in human oocytes turns out to be much less depressed by the addition of the cryoprotectants (about 6.5 °C compared with 23 °C for the mouse oocytes), indicating, first, a very efficient heterogeneous nucleation for these cells, and, second, the limitedness of the animal models in the development of the cryoconservation procedures for the human cells. Increasing the seeding temperature was found to provide a narrow window for the human oocytes to dehydrate without incurring lethal IIF.

Sensitivity of oocytes and embryos at the different development stages (oocytes, zigotes, two-, four-, and eight-cell embryos) to the cryopreservation was studied by Pfaff et al. The authors measured water and DMSO membrane permeabilities and found significant differences for eight-cell embryos only. They also found that the presence of DMSO increases the hydraulic permeability of the membrane. This effect was attributed to the amphiphilic properties of this cryoprotectant. The authors stress the need to carefully estimate the cell volume and surface, comparing two approaches: these parameters could be determined either from consideration of the entire embryo as a single osmotic entity or by treating it as a composition of the individual blastomers, with the correct account of the overlaps that occur due to cell impingement upon one another and the resulting deviation from the sphericity. To summarize the cooling rate effect on the cell survival, the following cases are possible:

1. $\tau_D < \tau_T$: excessive dehydration of the cell, "*solution effect*"
2. $\tau_D \approx \tau_T$: an intra/extra cellular equilibrium is established; optimal for the cell survival
3. $\tau_D > \tau_T$: IIF
4. $\tau_D \gg \tau_T$: vitrification.

Cells also differ in the sensitivity to the thawing rate, and this dependence is more sharp in the presence of the cryoprotective agent; as a rule, the more fast thawing is preferable. Osmotic volume excursions – shrinkage of cells during freezing and their re-expansion during thawing or during addition/removing of the CPAs – could result in the damage of some types of cells.

Cryoprotective Agents

The survival signature in cryopreservation could be modified with the use of the cryoprotective agents (CPAs). The presence of CPA

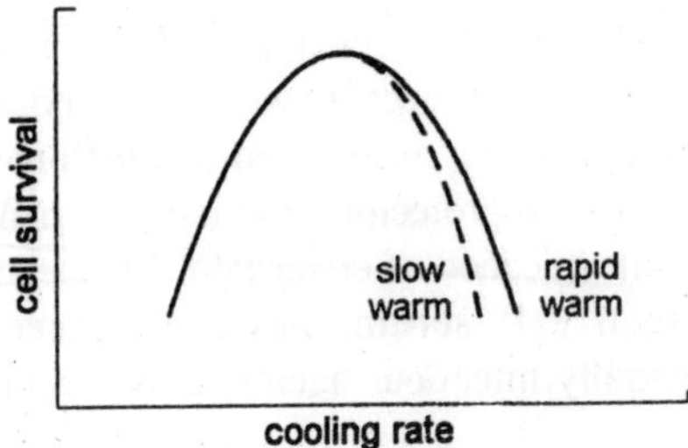

Fig. 4.3. Cell survival signature: effect of thawing rate.

increases the survival of the cells and shifts the value of the optimal cooling rate to slower cooling.

CPAs could be divided into two groups depending on their ability to penetrate the plasma membrane. The former (intracellular, or endocellular, or permeating such as propylene glycol (propane-1, 2-diol), ethylene glycol ($C_2H_4(OH)_2$, EG), glycerol ($C_3H_5(OH)_3$), dimethyl sulfoxid) reduces the concentration of the electrolytes and prevents cell shrinking in the hypertonic solution. If the intracellular cryoprotectant cannot diffuse out of the cell quickly enough during thawing to prevent the excessive influx of the free water and swelling of the cell, osmotic shock may occur resulting in cell rupture due to the extremely low intrinsic elasticity of the lipid bilayers. Rehydration at high temperatures, at which the membrane are in the fluid state, is less dangerous than at low temperatures at which the gel state of the membrane is observed.

The extracellular (exocellular, nonpermeating) cryoprotectors (sucrose, trehalose, d-mannitol, lactose) activate the water loss by the cell and decrease the probability of the IIF in the fast cooling process. Some cryoprotectors (glycerol, saccharides) prevent the partial

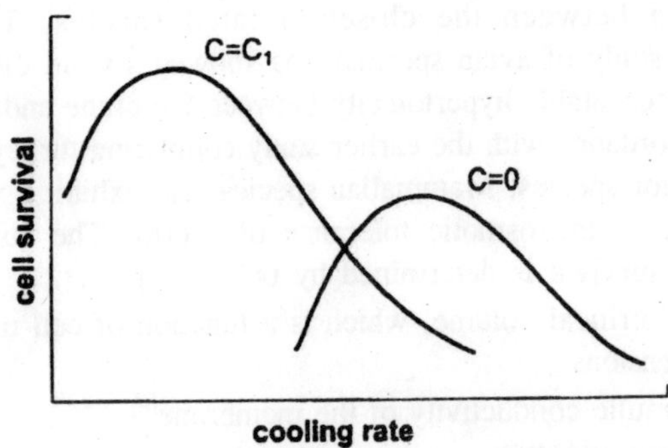

Fig. 4.4. Cell survival signature effect of thawing rate.

denaturation of proteins in the freezing solution and the eutectic crystallization. Both the low-molecular-weight compounds (dimethylsulfoxid, glycerol, disaccharides, and others) and high-molecular-weight polymers could serve as cryoprotectors; the metallic nanoparticles also could be used for the modification of cryoaction. Frequently conservation solutions contain fetal calf serum; however, there is a risk of transmission of potentially infectious agents, thus serum-free solutions are preferable now.

Cells placed in CPA solution are losing intracellular water due to the hyper- tonic character of the solution [688], and the cell volume decreases. After initial shrinkage, the water and CPA enter the cells to recover equilibrium and the cells swell. The reverse behavior is observed during the removal of CPA: initial cell swelling in response to the hypotonic extracellular environment is followed by shrinkage when water and CPA leave the cell. If the cell volume excursions are too large, irreversible membrane injury occurs. For example, osmotic cell shrinkage could cause an irreversible endocytic vesiculation that results in lysis during osmotic expansion, since new membrane material is not available rapidly enough.

One of the parameters that determine the quality of the cryoprotective agent is the number of water molecules it could bound: for example, one molecule of propylene glycole is surrounded by over 30 molecules of water.

Glycerol is usually used for mammalian species; ethylene glycol, DMSO for avian; glycerol, DMSO, methanol for fish species; and EG (due to its low molecular weight, high permeating ability; and low toxicity) for oocytes and embryos.

Efficiency of the CPA to increase the cell osmotic tolerance varies greatly even between the closely related species. The recent comparative study of avian spermatozoa showed sixfold difference in the extreme acceptable hypertonicity between the crane and the turkey sperm in accordance with the earlier study comparing turkey, chicken, and four raptor species. Mammalian species also exhibit a wide range of variability in the osmotic tolerance of sperm. The authors state that the cell survival is determined by two factors viz.:

1. The final critical volume, which is a function of cell morphology and dimensions
2. The hydraulic conductivity of the membrane.

Species-specific differences were found to be especially significant with respect to the toxicity of the CPA during the long-term storage in isotonic media. The effect of the CPA is related to its interaction with the cell membrane, which markedly alters the transition behavior of the membrane components and their protein kinetics. The molecules of the cryoprotective agent start to interact with the cellular membrane at the room temperature, changing the conformations of the integral proteins in the lipid bilayer. The cryoprotective properties of DMSO are attributed to the electrostatic interaction between its polar sulfoxide moiety with phospholipids of the membrane; at higher temperatures a hydrophobic association between DMSO and lipid bilayer is also significant. The cryoprotective properties of DMSO - highly polar aprotic solvent that mixes with water at ambient conditions in all proportions - are also attributed to the exposed sulfonyl oxygen that acts as a strong hydrogen bond acceptor.

The presence of the cryoprotector decreases the membrane fluidity and the temperature of the main phase transition as well as the temperature of the transition from the lamellar liquid crystal phase to the inverse hexagonal one, changes the hydraulic permeability of the membrane and its temperature dependence (the activation energy), the thermophysical properties of the solution (including the depression of the crystallization temperature), the morphology of the growing crystals: for example, the presence of DMSO activates the growth of the secondary dendrites; the cryoprotectors reduce the temperature at which the formation of the intracellular ice is being registered. Simple relation for the determination of the melting point depression and the eutectic temperatures were presented by Pegg for the two most popular CPA solutions: mixtures glycerol–NaCl–water and DMSO–NaCl–water.

Small solutes such as sugar remain in the very narrow spaces (at approximately 20% of water, the average separation between the membranes is about 1 nm) between the membranes at low hydration resulting from either freezing or, in problems of anhydrobiology, from desiccation.

Oligosaccharides and other low molecular weight compounds prevent too close convergence of the membranes when strong dehydration occurs, which delays their transition from the liquid to the gel phase.

The effect of the cryoprotector nature (sorbitol, DMSO, trehalose, saccharose) on the freezing of the lamellar phase of the model lipid membrane was studied by Yong et al. using nuclear magnetic resonance (NMR). This method allows to measure a fraction of the unfrozen water; heavy water D_2O was used as the solvent in the cited paper.

The severe cell dehydration (the water content reduces to about 10%) results, in addition to the increase of the intracellular solution concentration, to the contact of the nonaqueous intracellular components, including membranes. When these components are in close proximity (the gap could be as small as 1 nm or even less), the repulsive hydration forces that are characterized by the exponential decay with the distance will reach extremely high values that can lead to membrane deformation. In these conditions, stacks of membrane often are observed that resemble lamellar phases. The role of CPAs is to disrupt the water structure and thus to reduce the repulsive hydration interaction between the membranes.

It was shown that all the chemicals considered increase the hydration of the lamellar phase. However, it was found that pure osmotic effect dominates in the trehalose and saccharose solutions of low concentration. The authors proposed the thermodynamic model of the system in question that neglects the size of the molecules of water and cryoprotector compared with the distance between the lipid layers. The effect of solutes in the lamellar phases is separated into two components: a purely osmotic effect that is approximately similar for different solutes and specific effects that are determined by the peculiarities of the interaction of the given solute with water and with lipids.

When water is in equilibrium with ice and lamellar phase that contains no solutes, one write for the chemical potentials of water μ_w and ice μ_i

$$\mu_i = \mu_w = \mu_w^0 + P\nu_w,$$

where μ_w^0 is the standard chemical potential of water and ν_w is the partial molecular volume of water that is assumed to be equal to its bulk value. The mechanical equilibrium requires the interlamellar force per unit area $F = -P$. The experimental data could be fitted to the exponential decay for the force as

$$F = F_0 e^{-R/Rc},$$

where R is the number of water molecules per lipid molecule, R_c is the characteristic value, and F_0 is the extrapolated force per unit area at contact or using a characteristic length λ as

$$F = F_0 e^{-y/\lambda}.$$

The forces observed in the experiments are rather high: a value of the preexponential factor F_0 up to 4×10^9 N m^{-2} have been reported with typical decay length λ about 0.2–0.3 nm.

Assuming that at low hydration the interlamellar force is dominated by the exponentially decreasing hydration force, Yoon et al. obtained the relation for the chemical potential of the water in the form

$$\mu_w = \mu_w^0 + \kappa T \ln \gamma X_w - (F_0 e^{-R/R_c})v_w,$$

where the activity of water is represented as the product of the number fraction of water X_w and the activity coefficient γ. The presence of any solute independently of its nature will decrease X_w which, in turn, will result in the increase in the interbilayer separation y – this is called the purely osmotic effect. The nature of the solute could manifest itself in its effects on the activity coefficient, characteristic length, and the force at zero separation. These effects are called by the authors specific effects. The authors also assume that the interlamellar layer is in liquid state and note that if it is vitrified, and the model should account for the ability of the interlamellar layer to support the lateral stress. The lateral stress is important since it could influence the temperatures of the phase transitions of the lipid bilayers: membrane lipids in the gel phase have a lower surface area per molecule than those in the fluid phase, thus the compression favors the fluid–gel transition. The estimate of the transition temperature shift ΔT_m could be obtained from the Clausius–Clapeyron equation as

$$\Delta T_m = \frac{T_0 \Delta a}{2L}\pi,$$

where T_0 is the transition temperature at full hydration, Δa is the difference of the lipid surface area between the fluid and gel phases, L is the enthalpy of the phase transition, and π is the lateral stress in the bilayer.

If vitrification occurs in the bulk solution, it has little effect on the state of the membranes. However, if a vitrified layer is formed between the two fluid phase membranes during freezing, it greatly increases the stability of the fluid membrane phase and depresses the transition temperature of the main phase transition by more than 20 °C.

Trehalose provides significantly more effective protection of biomolecules then having the same chemical formulae $C_{12}H_{22}O_{11}$ maltose and sucrose due to the different configuration. Among these compounds, trehalose has the smallest partial molar volume, indicating a more compact conformational arrangement, especially at low temperatures, the highest glass transition temperature characterized by the highest values of the interaction strength parameter and of hydration number and thus a greater efficiency in destroying the tetrahedral network of water compatible with the structure of ice. Stabilization action of trehalose on the biological membranes is attributed to the interaction of the trehalose hydrogens with the polar head groups of the lipid macromolecules, resulting in the replacement of the water molecules at the membrane–solution interface.

Antifreeze proteins manifest their presence in two aspects. On the one hand, these chemicals inhibit the growth of ice crystals and ice recrystallization during thawing. AFP are extremely effective at inhibiting ice recrystallization even at very small concentrations (less than 1 μg ml^{-1}).

The molecules of AFP are incorporated into the growth steps on the surface of the ice crystal, splitting the crystallization front into the numerous segments; the increase of the front curvature results in the increase of the free surface energy and front propagation slowing; a significant part of the protein surface participate in this interaction. Using a scanning tunnel microscope one can see a system of grooves – the sites of the protein adsorption. The depth of these grooves – about 2–10 nm – allows to estimate the curvature of the crystallization front. MD computations show that AFP increases the energetic barrier for the homogeneous nucleation.

On the other hand, these proteins change the morphology of the growing crystal (or, in other words, modify growth habit) of the extracellular ice, increasing the cell destruction by the ice crystals that have spicular form. This property is exploited in cryosurgery: when AFP concentration is high, needle-like (bi-piramidal) ice crystals grow and mechanically destroy the cellular membrane. The AFPs could be injected into the tumor region prior to freezing to increase the destruction of malignant tissues. The micronsize highly destructive spicular shape of ice crystals is observed independently of the thermal conditions if AFP concentration is higher than about 5 mg ml^{-1}.

Crystals formed by two hollow hexagonal pyramids could grow in pure water if the supercooling is greater than 2.7 °C. The temperature of the ice–water phase boundary T_i could be written as

$$T_i = T_m(1 - d_1(\theta,\phi)\kappa_1 - d_2(\theta,\phi)\kappa_2 - \beta(\theta,\phi,V_n)),$$

where T_m is the melting temperature, θ, ϕ are the boundary orientation with respect to the two orthogonal axes, d_1, d_2 and κ_1, κ_2 are the corresponding values of the capillary length and the interface curvature, V_n is the normal velocity of the crystallization front. The kinetic supercooling $\beta(\theta, \phi, V_n)$ describes the deviation from the equilibrium for the given orientation and normal velocity. The variation of the crystallization front morphology is determined by the dependencies of the surface tension and the kinetic supercooling coefficient on the crystallographic surface orientation. More complex crystal shapes are observed for the higher supercooling exceeding 5.5 °C.

The "*growth-melt asymmetry*" observed for pure ice under pressure of 2,000 bar was reported for the crystallization-melting under the normal conditions in the presence of spruce budworm *Choristoneura fumiferana* antifreeze protein (the growth and melt shapes show facetted hexagonal morphology, but they are rotated 30° relative to one another) and attributed to the different adsorption of AFPs to the different crystallographic planes of ice.

Cell Interaction with Crystallization Front

The cells in the freezing solution could be mechanically damaged when they get trapped in the unfrozen solution channels between growing finger-like ice crystals. An experimental procedure to simulate the action of the ice crystals on the isolated cells in the freezing suspense was developed by Takamatsu. The author used two parallel plates to study viability of the deformed cells – the degree of the membrane deformation that causes the cell destruction – and found that viability of the considered primary prostatic adenocarcinoma cell (20 μm in the mean diameter d_0) drops when the gap between the plates h is about 6 μm that corresponds to 30% of the original cell diameter. The expansion rate of the cell surface is calculated assuming invariability of cell volume and written as

$$\gamma = \frac{A}{A_0} = \left(-\frac{\pi^2}{16} + \frac{2}{3}\right)\left(\frac{h}{d_0}\right)^2 + \frac{1}{3}\left(\frac{d_0}{h}\right) + \frac{\pi}{4}\left(\frac{h}{d_0}\right)^2 \sqrt{\frac{\pi^2}{16} + \frac{2}{3}\left(\left(\frac{d_0}{h}\right)^3 - 1\right)}$$

where A is the area of the cell surface and $A_0 = \pi d_0^2$ is that of underformed cell. Assuming uniform expansion of the membrane and free slip at the plate–cell boundary, the authors defined the critical expansion rate γ_{crit} – the tolerable limit of expansion found from experiments to be $\gamma_{crit} = 1.5$. A comment should be made on authors' terminology: it seems that the word "rate" be better omitted since the stationary state of the deformed membrane is considered.

The mechanical deformation of the plant cell wall depends on the extent of the water loss. The hydration ratio – the ratio of mass of unfrozen water in the cell wall to the dry mass – could be measured using quantitative NMR.

The crystallization front propagating in the supercooled solution is unstable. The first studies of interaction of the particle with advancing ice crystallization front were performed half a century ago. The authors of the former paper considered a thin layer of water between the particle and ice (*premelted film*). This layer structure is determined by the long-range intermolecular interactions (van der Waals or electrostatic) between ice and the particle surface.

Premelting between grains was considered as the mechanism of the ice crystals sintering in the middle of the nineteenth century by Tyndell (1858) and Faraday (1860) as a process that combines surface melting and refreezing called the thermal regelation. Gilpin was the first to model the thermal regelation.

Peppin et al. performed a linear stability analysis of a plane crystallization front during unidirectional solidification of a cell colloidal suspension. A suspension of bulk volume fraction ϕ_∞ is enclosed between two parallel plates with constant temperatures. The densities and thermal properties of ice, water, and particles are assumed constant and the cells are assumed to be the rigid spherical particles.

The volume fraction of particles ϕ in the coordinate system bonded to the crystallization front is governed by the mass conservation equation

$$\frac{\partial \phi}{\partial t} - V\frac{\partial \phi}{\partial z} = \frac{\partial}{\partial z}\left(D(\phi)\frac{\partial \phi}{\partial z}\right),$$

where V is the crystallization front velocity, $D(\phi) = D_0\hat{D}(\phi)$ is the diffusion coefficient, $D_0 = \kappa T/6\pi R\mu$, R is the particle radius, and μ is the fluid dynamic viscosity,

$$\hat{D}(\phi) = (1-\phi)^6 \frac{d(\phi Z)}{d\phi},$$

the compressibility parameter for the rigid sphere model $Z(\phi)$ is the fourth order rational function:

$$Z(\phi) = \frac{1 + a_1\phi + a_2\phi^2 + a_3\phi^3 + a_4\phi^4}{1-\phi/\phi_p},$$

$\phi_p = 0.64$ is the volume fraction at random close packing, the values of the coefficients $a_i (i = 1 - 4)$ could be found in the cited paper. Note that dependence of the diffusion coefficient D on the volume fraction ϕ is nonmonotonic.

Diffusion of particles in the solid phase is neglected. The mass conservation condition at the front is written as

$$(\phi_l - \phi_s)V_n = -D(\phi_l)\nabla\phi_l \cdot n,$$

where V_n is the normal front velocity. Concentrations in the solution ϕ_l and in the solid phase ϕ_s are related by the segregation coefficient κ_s:

$$\phi_s = \kappa_s\phi_l.$$

The boundary condition on the concentration far from interface reads as

$$\phi \rightarrow \phi_\infty \text{ at } z \rightarrow \infty.$$

The authors considered the case of slow freezing so that the segregation coefficient is small and approximately constant. The so called "frozen-temperature approximation"

$$T(z) = T_i + G_T Z$$

was used, where T_i is the temperature at the stationary plane interface and G_T is the constant temperature gradient. This approximation is valid if, in addition to the constancy of the thermal properties, the smallness of the latent heat of fusion of ice can be assumed.

In equilibrium the temperature at the planar interface is equal to the thermodynamic freezing temperature T_f, which for a hard-sphere suspension is

$$T_i = T_f(\phi_i) = \frac{T_m}{1 + m\phi_i Z(\phi_i)},$$

where $\phi_i = \phi_\infty/\kappa_s$ is the volume fraction of particles at the planar interface, T_m is the freezing temperature of the pure water, and $m = \kappa T_m/\upsilon_p \rho L$, $\upsilon_p = (4/3)\pi R^3$ is the volume of the particle, ρ is the density, and L is the latent heat of fusion of ice.

The temperature at the deformed interface is

$$T_i = T_f(\phi_i) + \frac{\gamma T_m}{\rho L} K$$

where ϕ_i is the volume fraction at the deformed interface, $K = h_{xx}(1 + h_x^2)^{-3/2}$ is twice the mean curvature of the interface, and γ is the interfacial surface energy. The second term on the right-hand side of the last equation represents the Gibbs–Thompson effect, that is the change in the melting temperature at the interface due to the curvature there. If this equation is to be considered macroscopically, the curvature of the interface should be much larger than the radius of the particle $|K^{-1}| \gg R$. The authors assumed that γ = Const, that is, surface energy does depend on neither the concentration of the particles nor the crystallographic orientation of the ice surface.

Linear stability analysis is performed by perturbing the front position h and the concentration C: $h = \bar{h} + h'$, $C = \bar{C} + C'$, where $\bar{h} = 0$ and $\bar{C}$ are solutions to the steady planar case. The perturbations h' and C' are sought in the form of normal modes

$$[h', C'] = [h_1, c_1(z)]e^{\sigma t + i\alpha x},$$

where σ is the growth rate of the perturbation and α is the wavenumber of the normal mode along the interface. The authors solved the characteristic equation $R(\sigma) = 0$ analytically for the dilute limit $C \rightarrow 0$ and numerically for the general case. The mechanism of the instability is essentially the same as for the dilute binary alloy considered by Mullins and Sekerka: a perturbation of the interface shifts a portion of the ice into the suspension; in the absence of the supercooling such a pertubation would melt; however, it will grow further into the supercooled region.

Linear stability analysis shows that increase of the particle radius stabilizes (through the increase of the surface energy term) the crystallization front, while the growth of the concentration has a destabilizing effect.

The cells behavior in the respect to the crystallization front depends on the velocity of the latter: the cell are forced from the phase boundary if the front velocity exceeds some threshold value V_c and are trapped by the ice for smaller velocities. Tada et al. performed an experimental study of this phenomenon using yeast cells for a wide range of the crystallization front velocity - from 0.8 to 80 μm s^{-1} - and found satisfactory agreement with the theoretical predictions. Chernov et al. derived an approximation for the critical velocity, assuming that the crystallization front could be presented by either the paraboloid or the union of the truncated paraboloid and the plane. These authors also studied the effect of the thermal properties of the particle and found that rejection is enhanced if the thermal conductivity of the particle is lesser than that of the ice.

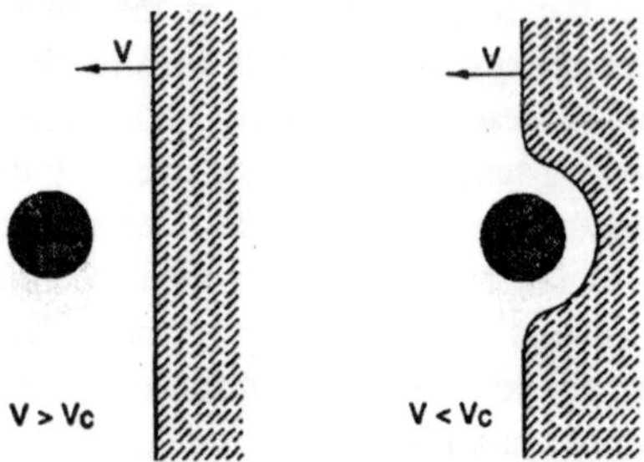

Fig. 4.5. Cell interaction with the crystallization front.

In addition to the particle radius, there are two linear scales in the problem. The first one is

$$\Gamma = \frac{T_m \sigma^{sl}}{\rho_s LRG},$$

(σ^{sl} is the surface energy, ρ_s is the density of the solid phase, G is the temperature gradient) which describes the effect of the crystallization front curvature, the second scale is related to the scale λ and to the character of the intermolecular interactions

$$l = \left(\frac{\lambda^{\nu} T_m}{G} \right)^{\frac{1}{(\nu+1)}},$$

$\nu = 3$ for Van der Waals forces, $\nu = 2$ for the electrostatic interaction. The critical radius R_c is determined by the condition $\Gamma = l$:

$$R_c = \frac{\sigma_{sl}}{\rho_s L} \left(\frac{T_m}{\lambda G} \right)^{\frac{\nu}{\nu+1}}.$$

For the case of large particles $R \gg R_c$, an analytical solution exists that predicts that the critical velocity is inversely proportional to the particle radius $V_c \propto R^{-1}$ and only weakly depends on the temperature gradient $V_c \propto G^{1/4}$ for Van der Waals forces; in the opposite case of small particles $R \ll R_c$, V_c does not depend on the temperature and $V_c \propto R^{-4/3}$.

For the case of the nonretarded van der Waals interactions being dominant ($\nu = 3$), the force per unit area that repels the particle from the crystallization front is written as

$$P_T = \frac{\rho_s L \lambda^3}{d^3} \equiv \frac{A_{sw}}{6\pi d^3},$$

where d is the distance from the particle surface to the crystallization front; the effective Hamaker constant A_{sw} is introduced into the second equality. This constant for interactions between particle surface and the solid could be determined experimentally or estimated using the dielectric response functions for the solid, the liquid, and the particle.

The net thermomolecular force on the particle is given by integral of P_T over the interfacial surface and for a spherical particle is

$$F_T = 2\pi R^2 \int_0^{\theta_c} \sin\theta \cos\theta P_T d\theta = 2\pi R^2 \rho_s L \frac{G l^4}{T_m} \int_0^{\theta_c} \frac{\sin\theta \cos\theta}{d^3} d\theta \quad \text{...(18)}$$

The distance from the particle surface to the crystallization front d is clearly determined by the angular position θ. Since the film thickness rapidly increases with θ and there is a cube of d under the integral sign, the net force is dominated by the region near the base of the particle and the precise choice of the upper integration limit θ_c is not critical.

This force is balanced by the friction force due to the viscous fluid flow. The so-called lubrication approximation could be used to relate the gradient of the hydrodynamic pressure P_1 in the liquid film to the volume flux of the fluid through the film near the particle base, which in turn is a function of the particle velocity U. This approximation is constructed accounting for the slowness of the fluid motion (the Reynolds number $Re \ll 1$) that allows one to neglect the inertial (convective) terms in the Navier–Stokes equations, getting linear Stokes equations. This force is written as

$$F_\mu = 2\pi\mu R^4 U \int_0^{\theta_c} \sin\theta\cos\theta \left(\int_{\theta_c}^{0} \frac{\sin\phi}{d^3} d\phi \right) d\theta. \qquad \text{...(19)}$$

Rempel and Woster introduced a small parameter ε_s as

$$\varepsilon_s \equiv \left(\frac{l^4}{\Gamma R^3} \right)^{1/3}$$

Then the film thickness near the base of the particle is of the order $\varepsilon_s R$ and the scale of the particle velocity

$$W = \frac{\rho L l^4 G}{6\mu T_m R^2 \varepsilon_s} = \left(\frac{\sigma_{s1} A_{SW}^2}{6^5 \pi^2 \mu^3 R^4} \right)^{1/3}$$

does not depend on the thermal gradient. The velocity scale is proportional to $R^{-4/3}$; thus, critical velocity will be lower for larger particles and they will be more readily trapped within the growing solid domain.

Vitrification

Experiments show that further increase of the cooling rate again increases the cells survival. Dumont et al. varied this parameter in the range 5–30,000 °C min^{-1} and observed the monotonous increase of the fraction of the survived cells starting the cooling rate equal to 5,000 °C min^{-1}. The authors suggested two alternative explanations: (1) the ice crystals formed in the cells are too small to cause cell damage (the characteristic crystal size decreases with the cooling rate); (2) vitrification occurs. Shimada and Asahina reported that formation of the intracellular ice crystals smaller than a critical size is tolerable in the rapidly frozen cells. This exact critical diameter was not cited, but it is believed to be less than 0.05 μm. Some protecting effect of the intracellular ice could be related to preventing cell dehydration during the slow cooling.

Vitrification is based on the notion that the homogeneous nucleation temperature (depending on the pressure, being equal to approximately

-39 °C at atmospheric pressure with a minimum -92 °C at 2 kbar) is not a thermodynamic constraint, just a kinetic restriction and thus not an absolute but a practical limit depending on the cooling rate. The rate of nucleation in water is a nonmonotonic function of the temperature due to the competition between the driving force for nucleation, that is, increases with supercooling and viscous slow-down of the water molecules mobility with the temperature decrease. Vitrification occurs if the temperature interval within which nucleation occurs can be bypassed rapidly.

Vitrification could be achieved by either of the following approaches (or their combination):

1. A superfast cooling (for the tiny volumes about 1 nl using liquid nitrogen supercooled down to -200 °C, the cooling rate of 130,000 °C min^{-1} has been demonstrated; computations show the possibility to reach the cooling rate over 300,000 °C min^{-1} using the specially designed cooling systems with increased heat transfer coefficients).
2. Use of the highly concentrated (up to 50% by mass) solutions of the permeating glass-inducing cryoprotective agents. The primary role of these additives is increase of the solution viscosity. Higher concentrations of CPA are needed for extracellular vitrification than for intracellular one since cells naturally contain high concentration of proteins that are beneficial for vitrification.

In the latter case the coexistence of the domains of crystalline and amorphous ice in the same sample is possible. It could be also observed in the conventional cryopreservation by slow control freezing of cells: a part of the system vitrifies due to the increase of the solutes concentration up to the point when the concentration of the residual solution is so high that it could vitrify in the presence of ice. The vitrification inside cells is assisted by the presence of the intracellular macromolecules.

Vitrification of the pure water to the HDA form could be achieved using significantly smaller cooling rates under moderately raised pressures (about 0.5kPa).

Usually, the traditional slow cooling methods of CP are referred to as equilibrium cooling, and the rapid procedures (vitrification) as nonequilibrium one.

Sometimes a state usually referred to as *doubly unstable glass* is achieved. In this state the vitrous system contains ice nuclei and probability of vitrification is very high if the warming is not rapid enough.

One of the effects of the cryoprotector's (especially nonpermeating macromolecules such as polyethylene glycol or polyvinylpyrrolidone) presence in the solution is the increase of the solution viscosity that assists in the transition to the glassy state. The specimens should be sufficiently dehydrated to avoid the IIF during the rapid cooling.

In vitrification, the cells are dehydrated mainly before the cooling by exposure to the high concentration of cryoprotective agents. A principal benefit of vitrification is the elimination of the need of the search for the compromise cooling protocol for different types of cell present in the tissue sample.

The drawbacks of cryoprotectors are their high toxicity that manifests itself after the transplantation and the difficulty in the uniform tissue saturation in cryosurgery and in cryopreservation of large samples (the microinjection techniques could only meliorate the latter problem). Some of the components of cryoprotective solutions such as formamide are known as mutagens. The toxicity could have the chemical or osmotic nature; in addition, the intracellular cryoprotectors could directly interact with the intracellular organelles. The osmotic toxicity is mainly defined by the strength of the hydrogen bond of the polar part of the cryoprotector with water.

The relative toxicity of three CPA – glycerol, DMSO, and dimethylacetamide (DMA) – for fresh fowl spermatozoa was studied by Tselutin et al. Glycerol was found the least toxic and DMSO the most toxic substance. However, DMA was superior to others as a cryoprotectant.

The reduction of the toxicity by the minimization of the perturbations of the cytoplasm consists in using cryoprotectors mixtures, including the trehalose or glucose addition to the concentrated cryoprotector solutions or the preliminary saturation of the tissues in the ethylene glycol solution. Combination of CPAs may result in synergetic enhancement of cell survival.

One approach to the solution of the coupled problems of osmotic stress and toxicity caused by the high concentrations of CPA needed for vitrification is the multistage procedure of stepwise decreasing temperature and increasing the CPA concentration in such a way as to reduce toxicity but avoid crystallization at each stage. The samples are warmed rapidly and the CPA is removed by the stepwise dilution. This approach with 4–5 stages was successfully applied to vitrification of rabbit arteries and cornea and mouse embryos. It should be noted that this procedure is very slow and work consuming. An alternative is the use of nonpermeating CPA such as sucrose as an osmotic buffer to lessen the probability of the osmotic shock.

The effect of the type of the cryoprotector on the formation of the amorphous ice and its stability was studied by Boutron using erythrocytes as an example. The main parameter that determines the behavior of the vitrified sample during the storage is the glass transition temperature: above this temperature the sample is a fluid with extremely high viscosity in the metastable state; below this threshold the sample is the amorphous solid. A comparison of different models that predict the glass transition temperature for the mixture "water-trehalose" showed that the general approach to estimation of the mixture property Y_{mix} as the weighted average of the components properties Y_i

$$Y_{\text{mix}} = \left(\sum Y_i f_i\right) / \left(\sum f_i\right),$$

where f_i is the fraction in the corresponding units (mass, molar, or volume) does not give acceptable accuracy. The authors considered the Couchman- Karasz and Gordon–Tayor models for the binary solutions and their extension to the mixtures of cryoprotectors of the similar chemical nature (e.g., low molecular weights carbohydrates).

The water glass transition temperature has been for a long time assumed to be equal to 136 K. However, recent investigations indicate higher temperatures from 145 up to 160 or 165K.

The thawing should be conducted in such a regime that prevents devitrification - nucleation and growth of ice crystals. Nucleation of cubic I_c ice along with the ordinary hexagonal ice was observed in the high concentration solutions. The nucleation processes in the ethylene glycol solutions were considered using Johnson–Mel–Avrami–Kolmogorov model that describes the evolution of fraction x of the crystalline ice in time as $x = 1 - \exp(-(Kt)^n)$, where the parameter n depends on the crystal morphology. The authors proposed to write the reaction rate constant as $K \propto D(T)N^{2/3}$, where $D(T)$ is the diffusion coefficient of the water molecules at the temperature T, N is the density of the crystallization nuclei. Sometimes the stability of the amorphous state is characterized by the critical heating rate, above which there is insufficient time for a vitreous sample to crystallize before the melting temperature is reached. Vitrified materials, which may contain significant residual thermal stresses developed during cooling, may require an initial slow warming stage to relieve these stresses.

Karlsson in his analysis of the devitrification estimated the number of nuclei in the cell as an integral over the cell volume and over the time interval of the homogeneous nucleation rate, assuming that the heterogeneous nucleation is suppressed by the presence of CPAs, and represented the evolution of the radius of the crystal nucleated at the time moment τ using the diffusion-limited crystal growth model by MacFarlane and Fragoulis as

$$r(t;\tau)=\sqrt{\int_{\tau}^{t}\alpha^2 D_{\text{eff}}dt'},$$

where α is the dimensionless crystal growth parameter, D_{eff} is the effective water diffusivity in the depletion zone near the growing crystal, both temperature depending.

The risk of devitrification during thawing depends on the nature of tissue. For example, Wusterman et al. have found, using the same vitrification procedures and the same CPA mixtures, that devitrification is a serious problem for the rabbit arteries while it is not observed in the case of the rabbit cornea. The probability of devitrification could be reduced with the help of the ice blocking agents.

The cryomicroscope allows one to observe the formation of the intracellular ice, cell deformation due to dehydration, and interaction with the ice crystals and other above-mentioned phenomena. Ice initiation and propagation in plants could be monitored using infrared video thermography.

Differential scanning calorimeter (DSC) allows the measurement of the heat release in the system as a function of time and temperature and thus study such phenomena as ice formation. When this experimental approach is supplemented by the model of the cell dehydration, it is possible to study water transport as well and to extract from the experimental data (DSC thermograms), by fitting to the model of cell dehydration, the membrane permeability parameters – the hydraulic permeability and its activation energy. It is usually assumed that the suspension contains identical cells that behaves independently. The analysis of the tissue freezing is also possible by invoking the so-called Krogh cylinder model, which represents the tissue as a cylindrical cellular region embedded into the extracellular space. This space includes interstitial space and vascular space, ignoring the complex topology of the interstitial compartments. The equation that governs the evolution of the volume of the Krogh cylinder vs. the temperature is similar to that describing the cell volume evolution and is written as

$$\frac{dV}{dT}=\frac{L_pA_cR_gT}{B\upsilon_w}\left[\ln\frac{V_0-V_b}{(V_0-V_b)+\gamma_s\eta_s\upsilon_w}-\frac{L\upsilon_w\rho}{R_g}\left(\frac{1}{T_R}-\frac{1}{T}\right)\right],$$

where T_R is the reference temperature and the hydraulic permeability depends on the temperature according to (3). Under assumption of the uniformity of the region containing the Krogh cylinder, the model predictions do not depend on the length of the latter. It is also supposed that resistance of the interstitial spaces to water transport is negligible. One evident distinction of this model in comparison with that of the single isolated cell behavior is that the extracellular space is not assumed to be infinite.

The study of freezing of tissues with freeze-substitution microscopy is known to encounter difficulties when the fraction of the interstitial space in tissues is large (it could be up to 55% in some types of tumors); DSC approach is free from this problem.

Cryoaction In Vitro

The distance between the cells in the tissue is much smaller than that in the suspension; nevertheless, the basic freezing processes are similar. The main differences in the cell behaviour are as follows:

1. A greater cell stability with respect to the dehydration for the case of the closely packed cells in the tissue
2. The large extracellular crystals that are relatively harmless in cell suspensions may be lethal in tissues containing tightly packed cells – in contrast to the cells in suspension, cells embedded into the extracellular matrix could not move away from ice crystals
3. An effect of the cells contact on the initiation of IIF (in the tightly packed cells a slower cooling rate may cause IIF) and ice propagation.

The latter issue has been studied in detail. The possibility of the ice propagation through the membrane pores was proved experimentally. This result is confirmed by the measurements of the time delay of the intracellular ice appearance in the neighbor cells. Ice propagation could results in additional membrane damage.

The effect of the extracellular ice on the intracellular nucleation is confirmed by numerous measurements of the maximal supercooling achievable in the isolated cells and tissue samples that is turned out to be larger in the absence of the extracellular ice in most cases.

Probably, the ice propagation occurs through the channels formed by the special protein called aquaporin. This protein is found in the various organisms from plants and bacteria to the mammalians. Aquaporin in the membrane forms tetramers with a functionally independent channel in each monomer that is permeable for the water but not for the charged molecules and ions, while the fifth channel in the center of the tetramer is permeable for both water and charged particles. There is a family of aquaporin proteins; while the geometry of the channels only slightly varies between aquaporin family members, water permeability may differ by several orders of magnitude. There are two permeabilities that should be distinguished: osmotic permeability p_f and diffusion permeability p_d. Both parameters could be determined experimentally: p_f is measured by the application of the osmotic pressure difference, while to measure p_d one should use isotopic labeling, that is, to use deuterium oxide - heavy water. However, it should be noted that D_2O has a lower zero-point vibrational energy, leading to the stronger O–D bond and a stronger deuterium bond compared with the hydrogen bond, in addition to greater mass of D_2O; as a result, the self-diffusion coefficient of D_2O is lower than that of H_2O by about 20%.

The chemical blocking of the pores reduces but fails to stop completely the ice propagation in the cells in contact. The intensity of the process does depend on the temperature: the ice propagation is impossible if the temperature is higher than some threshold value. This effect is related to the crystallization temperature depression in the capillar. It should be stressed that analysis of the ice propagation in the pores having diameter 0.8–2.5 nm using the continuum media models could give only the qualitative description. As Mazur noted, if a pore connected two cells with ice being only in one of the them too narrow to allow ice propagation, another destructive process is possible: water influx into the cell containing ice crystals and it freezing there with the probable membrane rupture as a result (similarity with freezing of water in the inorganic porous media is evident).

The dominant mechanism of the IIF depends on the cell type and the properties of the membrane. Muldrew and McGann considered three possible mechanisms:

1. The homogeneous or heterogeneous nucleation in the cytoplasm inside the cell
2. Ice propagation through the pores in the cellular membrane
3. Ice propagation through the defects in the membrane caused by the electric field that arises due to the interactions of the ions contained in the solution with the crystallization front - double

electric layer is formed at the phase boundary leading to the potential difference between the liquid and solid phases - the "*freezing potential*"

The probability of the electric breakdown of the membrane increases with lowering the temperature due to the increase of the surface tension of the lipid bilayer. The problem of pore formation in the membrane was considered both as a continuum mechanics problem and using MD computations. The growth of the extracellular ice results in the change of the transmembrane potential and reorientation of the membrane proteins that possess the dipole moment and, thus, in the change of the membrane permeability.

It turned out that no single model could explain all the experimental data on the intracellular nucleation obtained for the different types of cells, the different cooling rates and different types of cryoprotectors. For example, the magnitude of the freezing potential is proportional to the velocity of the crystallization front that in turn depends on the cooling rate, while the occurrence of IIF in experiments does not reflect these dependencies.

Muldrew and McGann proposed a new mechanism of the membrane destruction related to the intense water flow caused by the pressure difference and its mechanical action (friction) on the lipid bilayer of the membrane called *osmotic rupture* hypothesis (recently Mazur and Kleinhans suggested a more adequate term for this effect - *osmotic poration*). The drift velocity of the water molecule υ is defined as $\upsilon = h/\Delta t$, where h is the width of the hydrophobic region of the membrane and Δt is the average transit time for the water molecule needed to cross this region. The friction force per water molecule is proportional to the drift velocity $F_f/n = f\upsilon$ that is determined using the total mass flux as $\upsilon = (J_w h)/n$, where n is the number of water molecules moving through the membrane in time Δt. Thus the total force acting on the membrane $F = fJ_w h$. Using the estimate of the friction coefficient f as an analog of the Einstein relation $f = \kappa T/D_w$, where D_w is the water diffusion coefficient, the authors determined the average pressure acting on the membrane:

$$P = \frac{\kappa T}{AD_W} J_w h,$$

where A is the area of the membrane surface that could be compared with the maximal admissible one.

The spontaneous symmetry breakdown is caused by local fluctuation of the tensile force in the membrane. The authors assumed that the distribution of the fluctuations of a given strength over the area of the membrane could be described by a normal distribution about some mean value and for a given force the probability that pressure will exceed the local breaking strength of the membrane is given by the integral of binomial distribution that can be approximated as

$$P_{p>P_c} = \frac{1}{1+\exp(-b(P-P_{\text{av}}))},$$

where P is the pressure imparted on the membrane, P_c is the critical pressure that will exceed the local breaking force of the membrane, P_{av} is the mean stress of the membrane, and b is a parameter related to the standard deviation σ: $b = 1.70991/\sigma$.

The probability that a given constant in time force exceeds the strength of the membrane is then given as

$$P_{\text{rupture}} = \left[\frac{1}{1+\exp(-b(P-P_{\text{av}}))}\right]\left(1-\exp\left(-\frac{t}{\tau}\right)\right),$$

where the time constant τ is reciprocal of the mean rate of the occurrence of fluctuation α: $\tau = 1/\alpha$.

The confirmation of this "osmotic rupture" hypothesis is contained. Beney et al. have studied water transfer across the membrane of the yeast (*Saccharomyces cerevisiae*) cell that has a very high hydraulic permeability. The authors considered processes caused by the instantaneous osmotic pressure variation with different final osmotic pressures and found that the water efflux intensity alone does not control the membrane injury; they assumed that the state of the membrane is determined by a set of parameters that includes the intensity of the water flow, the drop of the osmotic pressure, and its average level.

Other essential factors that determine the ice propagation are the interaction of the cell with the extracellular matrix and the water content in the tissue. The differences in the behavior of the isolated cell and the cell in the densily packed system were observed for the spherical (with a diameter from 70 to 140 μm) aggregate of cells around a charged polystirol particle. The authors do not discuss the role the cell interaction with the surface; however, it should be significant even for the isolated cell.

An analysis of the water and cryoprotector redistribution in the islands of the biological tissue containing thousands of cells showed that while the properties of the cellular membranes are the main factors, the effect of the interstitial diffusion should be taken into account.

Experiments on freezing tissue samples in vitro confirm the dependence of the crystal size on the cooling rate. Sanz et al. using pork muscle as an animal model found that the smallest ice crystals (located near the surface of the sample where the cooling rate is maximal) four times smaller in diameter than crystals in the centre. Under appropriate conditions a net of the ice crystals in the extracellular space is observed.

When water flows out of the individual cell within the tissue, it flows into spaces such as capillaries. As a consequence, these interior spaces expand to as much as sevenfold damaging the cellcell connections in the tissue.

Another possible mode of ice propagation in tissues is by travelling along the blood vessels [838]. The growth of ice crystals in small vessels (diameter of arterioles is about 10–15 μm, of capillars 4 μm) differs from that in the large vessels. Recent investigations of the crystallization temperature depression and ice propagation in the glass capillars showed the existence of the critical radius below which the ice propagation slows down or even stops. The limiting supercooling ΔT accounting for the front curvature and osmotic effect could be written as

$$\Delta T = \frac{2 v_s \sigma^{ls} T_m \cos\alpha}{Lr} + \frac{v_l \pi_l T_m}{L},$$

where v_s and v_l are the specific volumes of the solid and liquid phases, π_l is the osmotic pressure of the solution, r is the capillary radius, α is the angle between the interface boundary and the capillar wall that satisfies the Young relation

$$\sigma^{lb} = \sigma^{sb} + \sigma^{ls} \cos\alpha,$$

where $\kappa = 2 \cos \alpha / r$ is the front curvature.

Experiments in the capillars with the radius from 1.5 up to 87.5 μm revealed that the critical radius r_0 below which the ice propagation is impossible with a good accuracy corresponds to the minimal wavelength λ_0 in the Mullins–Sekerka analysis of the plane front stability $r_0 \approx \lambda_0/4$.

The authors warn, however, that a direct transfer of the results obtained in the glass capillars to the problems of cryobiology should be made with caution due to the differences in the values of the wetting angle and elasticity of the blood vessels.

Vitrification of the cells embedded in their natural matrix is more difficult than in the case of suspension, presumably due to more severe restriction on the cells' movement.

In spite of the significant progress – for example, restoring of the elastic and viscous-elastic properties of the walls of the human blood vessels was reported as well as recovery of the endothelial function after vitrification of cornea at -110 °C – cryopreservation of tissues and especially organs encounters serious difficulties. First, the presence of cells of the different types prevents the finding of the optimal cooling and freezing protocols. Second, it is impossible to provide the high uniformity of the temperature for a large sample. As a consequence, the thermoelastic stresses arise in the tissues that could lead to the formation of cracks. The latter destroy the tissue integrity and could also cause devitrification during thawing, serving as the heterogeneous nucleation sites since the filling of cracks decreases the surface energy. The danger of cracking in vitrification is greater than in cryosurgery due to both the more fragile amorphous state and the higher temperature gradients. Cracking during vitrification in the glycerol solutions was for the first time reported by Kroner and Luyet.

These problems force the further study of the alternative approach – slow cooling in the presence of cryoprotector. Since the presence of the cryoprotector could significantly change the thermal expansion coefficient, the nonuniform saturation of tissues with the cryoprotector also results in the appearance of the mechanical stresses; the extent of the deformation depends on the cryoprotector concentration and the cooling rate. An analysis of the stresses in the tissue sample in the case of the mixed crystalline-amorphous state requires accounting for the viscous-elastic properties of the latter.

Cryoaction In Vivo

Cryodestruction of the biological tissues in vivo turns out to be more severe than it could be predicted on the basis of the in vitro studies. The main additional factor of the damage is an injury to the blood circulation system: the most destruction is observed for a small vessels; as a rule, large vessels restore their properties after thawing. The first written account of the blood vessel's damage due to freezing was given in 1877. An experimental separation of the cellular and vascular cryodestruction was performed by Le Pivert.

The extent of the vascular damage that is based on the water redistribution is determined by the thermal history. One can distinguish three groups of damage effects:

1. The mechanical deformation of the vessel's walls due to the ice formation in the vessel (the observed increase of the diameter of the small vessels is more than twofold)

2. The death or the separation of the epithelian cells that cover the inner surface of the vessel and normally provide a nonthrombogenic blood-contacting surface and the permeability barrier between the blood and the vessel wall as well as serve as a signal transduction interface that regulate the vascular smooth muscle behavior
3. The damage of the blood circulation itself that is observed in several hours or days after the operation - vascular stasis sometimes referred to as *reperfusion injury*.

Vascular Stasis

The vascular injury caused by freezing was described by Lewis and Love who studied response in human skin during and after freezing to -5 °C in 1926. The vascular injury happens at much higher subzero temperature than the direct cell injury. After the complete thawing, blood flow returns to the frozen region accompanied by edema. The freezing preferentialy destroys the microvasculature responsible for the nutrient and oxygen delivery to the tissues, which leads to the cellular anoxia and hypoxia and, finally, results in necrosis. Damage to the vessel walls manifests itself as the frequently observed leakage through the vessel wall. Removal of the endothelial cells from the wall exposes the underlying thrombogenic connective tissues, leading to the platelet aggregation and the thrombus formation also frequently registered in experimental studies of the freezing effects. Small blood vessels become completely thrombosed in 3–4 h after thawing, while large could remain open up to 24 h. Edema developes and progresses over a few hours.

In addition to the mechanical damage to the endothelial cells due to the vessel distension and wall tears, free radical propagation and neutrophil activation are considered. The free radical formation could be attributed to the excessive oxygen delivery after thawing - hyperperfusion, since caused by the low temperatures vasoconstriction will be replaced, due to the compensatory regulation, by vasodilatation. There are indications that free radical inhibitors diminish the freezing damage. The injury connected to the neutrophil activation could be mediated by their attachment to the endothelium and release of the enzymes whose task is to digest dead and damage cells. Experimental intravital techniques showed the adhesion of neutrophils to the vessel wall. Neutrophil activity also could release free radicals. While many of the cells near the target frozen region are left unaffected, some cells that are highly sensitive to their ionic content such as platelets show the reaction to moderate cooling. Cooling of these cells to temperatures lower than their lipid transition temperatures cause calcium influx that, in turn, trigger platelets activation, leading to the cascade of events that would end in platelet aggregation, the obstruction of the blood vessels in the region around the frozen tissue.

The heat transfer contribution by the blood flow in cryosurgery problems is considerable, but not so important as in the hyperthermia procedures or in the analysis of the physical reactions to the burns. It was established in the experiment in vitro with the artificial blood circulation (emulation of the heat transfer in vivo) and in the combined numerical and experimental study of osteonecrosis that this mechanism of heat transfer could be significant during thawing. Effect of blood perfusion on heat transfer is reduced with each subsequent freeze-thaw cycle. By all means, a large blood vessel in the domain of the cryoacton or in its near vicinity is an exception. In this case the contribution of this vessel should be considered explicitly to correctly describe the zone of necrosis.

In contrast to the cryopresevation, the cooling rate in cryosurgery is not the principal parameter partly because of its great variation over the domain of cryoaction. The dependence of the local cooling rate (left) and the fraction of the destructed cells (right) are qualitatively illustrated by Fig. 4.6. The right curve is obtained from the left one by the transformation using the cell survival signature.

The slow thawing is important, resulting in, first, in the temporal hypothonic cells' environment (and, as a consequeness, water inflow into the cell and the probable disruption of the cellular membrane); second, in the intense ice recrystallization that is most significant in the temperature range [–40 °C, –15 °C]; finally, the slow thawing increases the effect of the oxidative stress.

The repeated freezing is important: it considerably increases the size of the zone of the cryonecrosis and critical temperature needed for the direct cryoablation: from –61.7 °C (95% confidence interval –

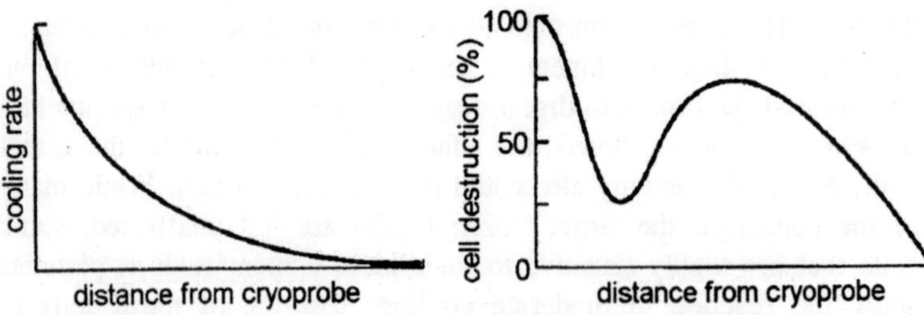

Fig. 4.6. The cooling rate (left) and cells destruction (right) vs. the distance from the cryoprobe.

74.5 °C to –48.9 °C) for the single freeze to –41.4 °C (95% confidence interval –49.9 °C to –33.0 °C) for the double freeze in experiments by Larson et al. One of the reasons of the repetition effect is the increase of the thermal conductivity of the tissue due to the cellular membranes breakdown during the first freeze–thaw cycle. The saturation of the effect of the multiple cycles is observed after 4–7 cycles. The intracellular ice crystals are larger in the second freezing cycle. Multiple cycles are especially important in the prostatic cryosurgery and other cases of close position of the tumor to the vital organs, permitting, via the reduction of the lethal temperature, a closer approach to the margins of the malignant tissue domain without endangering other organs. A longer interval between the cycles leaving the tissues in the hypothermic state allows time for the development of the microcirculation failure. On the other hand, as experinents with animals showed, multiple cycles could result in the animal death due to the cryoshock. The actual mechanism is poorly understood, but it is probably related to the releasing of the intracellular contents of many cells into the circulation, causing the response from the cells of the blood or vasculature or to the too great strain on kidneys.

5

Ice in Biological Preservation

The target of the action of low temperatures is the water contained in large amounts in the biological objects, thus the phase transitions of water are of primary importance. A typical cryosurgery operation lasts between several minutes and an hour. Cryoaction in medicine consists of a single or several (in cryosurgery) cycles "freezing–exposition–thawing". The interval between the cycles in cryosurgery considerably increases the destruction of the living tissues, leaving the tissue for a longer time in the hypotonic state and providing the sufficient time needed for the blood microcirculation failure to develop.

Cryoaction could be modulated in time (the so-called "*dynamic*" *cryosurgery* or *cryocycling*); as experiments show, the frequency of the temperature oscillations, caused by cryoprobe, is conserved throughout the entire region, while its amplitude decays with distance.

Amorphous Ice

The possibility of a fast transfer of the object to the glassy state without crystallization (*vitrification*, from Latin vitri – glass; B.J. Luyet used also the term "amorphization") is determined by the ability to attain the high enough cooling rate to pass through the ranges of the crystallization temperatures within the time interval shorter than that is needed for the formation of ice crystal nuclei, thus vitrification is easier to accomplish for a small object. Usually, vitrification of the biological objects is performed in the presence of the *cryoprotective agent* (CPA). Type of the cryoprotectant, its concentration, as well as the protocols of the tissue saturation and of a removal of the CPA influence the outcome of the cryoaction. Partial vitrification could occur during the conventional cryopreservation when the residual solution becomes so concentrated that could vitrify in the presence of ice.

Two phases of amorphous ice are known for a long time - the *low-density amorphous* ice (LDA), also called the *amorphous solid water* (ASW) or the *hyperquenched glassy water* (HGW), and the *high-density amorphous* ice (HDA); it is under discussion whether the very-high-density amorphous ice (VHDA) that could be obtained from the HDA by heating at the high pressure should be considered as the "true" independent phase. All three phases have a tetrahedral structure of the hydrogen bonds net, but the HDA (VHDA) has one (respectively, two) additional molecules between the first and the second shells of neighbors. Numerical experiments show the possibility of an existence of the fourth phase of amorphous ice.

Recent studies indicate that the HDA is a metastable phase while there are only two stable phases - the LDA and VHDA. A possible porous structure of the different phases of amorphous ice (and, hence, the ability to absorb and retain gases) is of great importance for physics of the extraterrestrial objects - the comets, the satellites of the outer planets, and the interstellar dust. The phase transitions from the vapour, the liquid, and the hexagonal ice I_h into the amorphous ice as well as the transitions between the different phases of the amorphous ice are considered in detail in the recent reviews by Debenetti and Loerting and Giovambattista. On the relation of the polyamorphism of water to the interpretation of the freezing phenomena in the solutions, see the paper by O. Mishima.

The term "order parameter" frequently used in description of the glassy state has a somewhat different meaning than in physics of critical phenomena and phase transitions; it is even proposed to use "order" parameter in the issues relevant to the glassy state to avoid confusion. The order parameters are related to the analysis of the rare transitions between the stable or metastable states in the free energy surface. Various activated events could be described in terms of the transitions between stable (global minimum) and metastable (local minima) basins in the free energy landscape separated by the free energy barriers (transition states) as a set of paths connecting the relevant basins. The order parameters are useful for the classification of the different metastable states accounting for their characteristics such as symmetries associated with different phases and understanding of the mechanisms of transitions analyzing the evolution of order parameters along the paths in the free energy landscape as well as of the equilibrium properties of the system dependence on the relation of the free energy to order parameters. The latter relation could be obtained in two ways:

1. By construction of phenomenological theories of phase transition (*top-down approach*); the free energy functional is usually constructed using symmetry considerations;
2. By the coarse graining (*bottom-up approach*) of the microscopic Hamiltonian.

"Physical" order parameter is indeterminate at the critical point (the point of the phase transition); transition to the glassy state is not a genuine phase transition, but a dynamic phenomenon and the glass transition temperature is not strictly defined. The meaning of the order parameters in the glassy state context is rather just a set of variables that define the properties of the glassy state. There is a discussion on how many such order parameters are sufficient for the complete description of glass, which was started in the middle of the 20th century with the appearance of the theory by R.O. Davies and G.O. Jones (that idealized the glass transition by treating it as a true phase transition) and is not finished yet. The question "how many" could be reformulated to the constraints on the so-called Prigogine-Defay ratio Π defines as

$$\Pi = \frac{\Delta c_p \Delta \kappa_T}{T_g (\Delta \alpha_p)^2},$$

where Δc_p is the difference between liquid and glass isobaric specific heat per unit volume at the glass transition temperature T_g, $\Delta \kappa_T$ is the liquid–glass difference of isothermal compressibilities, and $\Delta \alpha_p$ is the liquid–glass difference of isobaric thermal expansion coefficients.

It should be noted that the Prigogine–Defay ratio Π is not strictly defined since, first, the parameters entering this criterium are usually obtained by extrapolation to the glass transition temperature that is not strictly defined itself, and, second, the glass state relaxes continuously making its properties a function of time. One of the approaches to the description of glass transition consists in freezing-in of one or more order parameters related to the internal structural degrees of freedom involved in vitrification.

The inequality $\Pi \geq 1$ follows from the Davies–Jones theory, turning into the equality if there is exactly one order parameter; the case of the single parameter also implies that disturbances should decay to the equilibrium exponentially.

Experiments show that for the most glass-forming liquids, $2 < \Pi < 5$ and the relaxations almost always are nonexponentials, seemingly supporting the need for several order parameters. However, the recent study of this issue by N.P. Bailey et al. reported in the just cited paper (the authors analyzed the frequency-dependent thermoviscoelastic response functions of the metastable liquid) showed that properties of many liquids, in particularity those in which intermolecular interaction are dominated by the van der Waals forces, could be determined by the single order parameter (no claim is made that the complete description of the molecular structure is provided by this parameter), while the description of the properties of the hydrogen-bonding liquids such as water requires more parameters.

The process of the amorphous ice aging is of more interest for cryobiology, since it defines the conditions and longevity of the storage of the vitrified biological objects. The most important aspect of aging is the formation of critical embryo/nucleus in the metastable supercooled liquid and their extinction upon heating. Experiments on the supercooled organic glass-forming liquids showed that generation of crystal nuclei is possible at temperatures over a hundred degrees lower than the glass transition temperature.

The amorphous ice by the water vitrification was obtained a half century later than by the deposition from the vapour phase on the cold substrate at low pressure by the method resembling modern molecular beam epitaxy used to grow the semiconductor heterostructures. Crystallization in the biological objects differs greatly from the growth of the inorganic semiconductor crystals for a number of reasons: the anomalous properties of a water itself, the effect of the hydrophobic and hydrophilic biological surfaces on the water behavior, and the peculiarities of the crystallization in the heterogeneous media.

WATER

First of all, pure water in both liquid and solid phases (the last known phase of water, ice XII, was discovered in 1998) possess a number of anomalous properties; a complete list of them contains over 50 entries. The anomalies manifest themselves more strongly when the water is supercooled (the term *supercooled* is used more frequently than the more exact one *undercooled*).

Water has uncommon properties also as a solvent – for example, the mobility of the solvated ions nonmonotonically depends on the radius of the ion and do depend on the sign of the charge, in contrast to continuum theory of the ion behavior in the solution.

Probably the most known and certainly the most important water anomaly is that it is liquid at the normal conditions, in contrast to many similar small molecules. Water molecule itself is "an unremarkable small molecule". All water anomalous properties are defined by interactions between water molecules. Perhaps the simplest explanation is of its liquid state at the ambient temperature: unusually high strength of interaction between two water molecules (about 20 kJ mol^{-1} is that is an order of magnitude larger than the typical thermal fluctuation at room temperature ($\kappa T_{room} \approx 2$ kJ mol^{-1}). In contrast, many other small molecules interact via the much weaker van der Waals forces (the typical energy of the interaction is of the order of κT).

Most of the anomalies of the water originate from the ordering of the water molecules in the different states (the ice crystals, the liquid water, and the gas clathrates); this order, in turn, is just a manifestation of the cooperative effect of the structurally and dynamically nonuniform net of hydrogen bonds.

The thermodynamic water properties are determined first and foremost by the static inhomogeneities - a local structure, namely, the fourfold (in the average - computations using the different interaction potentials show a considerable number of the water molecules having the coordination number from 3 to 6) net coordination - the organization of the water molecules in tetrahedra. The angle between H–O bonds in the water molecule (HOH angle) is well established to be 104.52°, which is close to both the tetrahedral angle (circa 109.5°) and the internal angle of the pentagon (108°), providing thus the ability of water molecules, in addition to tetrahedral structures, to form near-planar pentagonal ring structures. The standard simple model of the water molecule having two regions of positive charge close to the positions of the hydrogen nuclei and two regions of negative charge ("lone pair electrons") tetrahedrally displaced with respect to the positive charges is, as high quality quantum mechanical computations show, an oversimplification. The negative (lone-pair) regions appear to be much closer to the molecular center than the positively charged regions to the hydrogen atoms and the overall charge distribution is near trigonal rather than tetragonal. The molecule's repulsive core is nearly spherical; however, deviation from the sphericity may be important in the assemblies of the water molecules. If a water molecule is surrounded by four hydrogen-bonded neighbors, in two of these interactions the central water molecules act as hydrogen-bond donor by directing its positively charged hydrogen regions to negatively charged regions (lone-pairs) of each of the two neighbor molecules. In other two interactions, the central molecule acts as a hydrogen bond acceptor of two neighbor water molecules pointing their hydrogen regions towards the lone-pairs of the central molecule. Water seems to be unique in having this donor:acceptor ratio equal to 2:2, which is considered as a very important attribute with respect to the chemical and biological consequences.

A cause of the tetrahedral near order is the sp^3 hybridization of the electron orbits. The dynamical properties - both translational and rotational - are traced down to the defects of the hydrogen bonds' net.

Both experiments and computer simulations show that in the supercooled liquids in some transient regions, relaxation times greatly differ from that in other fluid regions. In the early 1960s, G. Adam and J.H. Gibbs postulated the existence of cooperatively rearranging regions (CRR) whose molecules moves essentially independently of the rest of the system. M.G. Mazza et al. studied this issue in detail using MD simulation of the system comprising 1,728 water molecules interacting via the extended simple point charge potential (SPC/E model) for the temperature range from 350K down to 200 K. Computations have been used to extract a rotational mean square displacement

$$\langle \varphi^2(t) \rangle \equiv \frac{1}{N} \sum_l (\varphi_l(t) - \varphi_l(0))^2$$

and a rotational diffusion coefficient

$$D_R \equiv \lim_{t \to \infty} \frac{1}{4t} \langle \varphi^2(t) \rangle.$$

The computations showed that rotational dynamics is spatially heterogeneous. The study of the temporal evolution of the clusters containing more easily rotating molecules compared with that in the bulk (the authors called them *rotational heterogeneities* (RH)) and the radial distribution function of oxygen atoms within HR, which show the amplitude of the first peak that is greatly enhanced in comparison with bulk water, showed that there is a strong tendency for the more mobile molecules to be neighbors. These molecules, as a rule, have more than four neighbors and thus represent "*defects*" in the tetrahedral network of hydrogen bonds. The translational heterogeneities, while not identical to RH, certainly have similar features and the same nature.

The net structure changes significantly in the case of the spatial confinement – for example, for a quasi-one-dimensional water in the single-wall carbon nanotube (SWCN), the coordination number is less than 2. The ordering in the liquid phase is based on the structures similar to the hexagonal ice I_h – the water hexamers $(H_2O)_6$, which can form a number of configurations with close energy; the pentamers $(H_2O)_5$ observed in the clathrates are grouped into the dodecahedra. Dodecahedral water clusters have also been reported for protein surfaces. Using an elementary cell containing 14 molecules, an icosahedral structure containing 280 molecules could be constructed that can transform between the two configurations with the different density without breaking the hydrogen bonds. This model was developed by combining sheets of boat-form and chair-form water hexamers that are present in the crystallographic lattices of hexagonal and cubic ice.

Water displays a remarkable combination of rigidity and flexibility: while the hydrogen bonds network seems to be "*rigid*" with respect to the thermal fluctuations at normal temperatures, the large number of coordination defects and water's ability to vary its hydrogen bonding and participate in cooperative motions provide high molecular mobility and anomalously high proton conductivity in liquid water. The first explanation of the latter was suggested two centuries ago ("*Grotthuss mechanism*") using the hydroxonium ion H_3O^+ as a mediator agent in the proton transport; modern experimental evidence is in conflict with this simple scheme; mechanism of the proton conduction based on the ab initio computations involving the hydrogen-bond breaking process as a possible rate-controlling stage of diffusion is considered in detail by D. Marks et al. It should be noted that the high diffusivity of protons in the liquid water is relevant to some biomolecular process; also, it is the basis of the hydrogen energy – fuel cells that directly convert a fuel source energy to electricity heavily rely on this property.

In normal liquids, density and entropy fluctuations decrease with the temperature while in the water, due to the local tetrahedral geometry of the molecule interactions, these fluctuations increase. Quantitatively this behaviour could be described with the spatial (tendency of the molecules to adopt preferential separations) and orientational (deviation from the local tetrahedral arrangement) order parameters that both decrease as the volume is decreased largely due to the increased dispersion in both bond angles and bond lengths. The net of hydrogen bonds serves as a cornerstone for some water models: it is considered as a set of space filling curves; the temperature dependence of the water properties is expanded as a series in the structure functions that describe the equilibrium net of hydrogen bonds (the first function gives an average number of bonds per molecule, the second one describes the deviation from the tetrahedral structure, etc.).

The degree of the water ordering increases (respectively, decreases) when cosmotropic (chaotropic) ions are present. The character of the ion effect on the water depends on the ion radius and the charge sign and manifests itself, *inter alia*, in the viscosity η dependence on the concentration c

$$\frac{\eta(c)}{\eta_0} = 1 + A\sqrt{c} + Bc,$$

where the Jones–Doyle coefficient B is positive for the cosmotropic and negative for the chaotropic ions. On the experimental study of the effect of the polar, nonpolar, and charged molecules on the water structure see the paper by D.T. Bowron. The ordering of the water molecules increases the water thermal conductivity, making it close to the thermal conductivity of ice.

The *trans*-conformation of the hydrogen bond due to the additional interactions of protons with the nondivided electron pairs that does not participate in the forming of the bond in question proved to be stronger than the *cis*-conformation. As a rule, two OH groups of the same molecule participate in hydrogen bonds of different strength. A model of the continuous net of hydrogen bonds could account for both the strong and the weak bonds.

The presence of the far order in water is confirmed by the shape of the radial (pair) distribution function $g(\mathrm{r})$, found numerically in MD computations or experimentally by scattering of the neutrons or the X-rays. These methods are the best currently available means to probe the atomic structure of aqueous solutions. The distribution functions $g_{\alpha\beta}(r)$ of pairs of atoms α and β of the systems are derived using Fourier transformation from the neutron or X-ray scattering patterns obtained experimentally. Knowledge of these functions allows to determine atomic correlation distances, coordination numbers, and the extent of the local order around a particular atom type. An accuracy of the measurements could be enhanced using the so-called *isotopic substitution* to enrich the experimental information by exploiting the differences between the scattering patterns of isotopically labeled samples.

Both the distribution of atoms and of the interstitial voids are of interest. The *difference* distribution function $G(\mathrm{r}) = r^2\,(g(\mathrm{r}) - 1)$ is well suited for visualization of the large distance correlations. The orientation-correlation function $g(r,\Omega_1,\Omega_2)$ could not be found directly from measurements and requires post-processing by the method of empirical potential structure refinement (EPSR) that invokes the Monte-Carlo procedures. The total correlation function $g(r, \Omega_1, \Omega_2, \Omega_{12})$ includes the polar coordinates of the vector connecting the centers of the molecules. It is proposed in the just cited paper to analyze the coordination shells using the correlation functions for the spherical layers $R_{min} < r < R_{max}$. While the radial distribution function is routinely obtained in simulations, the determination of the orientational distribution function that depends on bond angles for each distance between molecules is rather labor-intensive, thus approximations to it are of great interest.

The water structure could be described with the aid of integral equation as was pioneered in 1914 by Dutch physicists Leonard Salomon Ornstein and Frederik Zernike. Their idea was to reformulate the equation for the total correlation function $h(r_{12}) = g(r_{12}) - 1$ (g is the radial distribution function), that is, the measure for the "influence" of molecule 1 on molecule 2 at a distance r_{12} by splitting this influence into two parts, direct and indirect. The direct one is given by the so-called direct correlation function $C(r_{12})$, while indirect contribution to the "influence" is weighted by the density and averaged over all the possible positions of particle 3.

Method of Ornstein–Zernike (OZ) integral equations is especially useful in combination with the so-called central force models within which the water is considered as a simple mixture of the oxygen and hydrogen atoms and no distinction is made between the atoms interactions within the same molecule and for the atoms belonging to the different molecules. One is thus relieved from the necessity to describe the orientational properties of the hydrogen bonds and to monitor the orientation of the water molecule. The OZ equation for the pair correlation function of the spatially homogeneous system reads as

$$\gamma(r_{12}) = h(r_{12}) - C(r_{12}) = \rho\int C(r_{13})h(r_2 3)\mathrm{d}r_3 \qquad ...(1)$$

where

$$h(r) = g(r) - 1 = \exp\left[-\frac{\Phi(r)}{\kappa T} + \omega(r)\right] - 1. \qquad ...(2)$$

Here $g(r)$ is the usual radial correlation function defined as

$$g_2(r_1,r_2) = V^2\iint\frac{1}{Q_N}\exp\left[-\frac{U_N(r_1,r_2,\ldots r_N)}{\kappa T}\right]\mathrm{d}r_3\ldots\mathrm{d}r_N, \qquad ...(3)$$

where Q_N is the configuration integral and U_N is the potential energy of the system, $\omega(r) = \omega(r;\ T,\ \rho)$ is thermal potential, and $\rho = V/N$ is the density. In the case of central forces, $g_2(r_1,r_2) = g(|r_2 - r_1|) = g(r)$.

The OZ equation (1) is exact. However, its analysis requires the introduction of approximations arising from the generally unknown relation $C = C(h(r))$ that could be reformulated as finding of the so-called bridge-potentials $B(r)$. The latter are expressed as an infinite series of the irreducible diagrams. The problem of finding approximation for the bridge-potential is especially complex if long-range electrostatic interactions are present in the system. If, however, an approximation for the radial correlation function is found, all macroscopic parameters could be relatively easily determined.

A recent review by Sarkisov considers both the phenomenological models of water and the models based on the integral equations for the correlation functions of the Ornstein–Zernike type.

The computations indicate that a first-order phase transition "liquid-liquid" is possible at low temperatures. This transition is analogous to the one observed in phosphorus that has a similar structure, metal-metalloid melts (Ni–P, Fe–P, Ni–B), some metal alloys (Ni–Zr), presumably, in solutions of some macromolecules, and some other fluids. The liquid–liquid transition is also predicted to occur at high pressure in BeF_2. This transition coincides with a violation of the Stokes–Einstein relation

$$D = \frac{\kappa T}{6\bar{\omega}\eta R}, \qquad \text{...(4)}$$

where D is the diffusion coefficient, η is the fluid viscosity, and R is the hydrodynamic radius of the particle. Since the determination of the viscosity from MD computations is not straightforward, the authors of the just cited paper used instead in their study the time τ_α of the so-called α-relaxation: this variable exhibits the same temperature dependence as η.

It is known that both the Stokes–Einstein relation (4) and its analog for the rotational diffusion - the Stokes-Einstein–Debye relation

$$D_r = \frac{\kappa T}{8\bar{\omega}\eta R^3}, \qquad \text{...(5)}$$

that were derived by the combination of classical hydrodynamics (Stokes law for flow around the sphere) and kinetic theory - hold for low-molecular-weight fluid for the temperatures significantly higher than the glass transition temperature T_g: $T \gg T_g$. When liquid is deeply supercooled, these relations greatly overestimate the diffusion coefficients.

M.G. Mazza et al. have established that origin of this breakup could be traced down to the dynamic heterogeneities in the supercooled liquid that does not become a glass in the spatially homogeneous fashion. The authors also found that degree of the violation of the Stokes–Einstein (4) and the Stokes–Einstein–Debay relation (5) varies with temperature, since following from these relations a temperature-independent ratio

$$\frac{D}{D_r} = \frac{3}{4R^2}$$

does change with supercooling.

In the *low-density liquid* (LDL) an open locally ice-like network of hydrogen bonds is present, while in the *high-density liquid* (HDL) the local tetrahedrally coordinated hydrogen bond network is not fully developed; the structures of LDL and HDL are similar to those of the corresponding phases of amorphous ice LDA and HDA.

The behavior of the disordered media under the temperature variation is usually attributed to their ability to form stable structured objects using the directional bonds. It is turned out, however, that the directional character of the bond is not the necessary feature: the numerous water anomalies could be reproduced assuming the tetrahedral (sp^3) coordination described by the spherically symmetric interaction potential with the soft core. Such potential (ramp potential with two characteristic lengths - an impenetrable hard core and a penetrable soft core) is called sometimes the Jagla potential and the corresponding fluid, displaying many water-like structural, dynamic, and thermodynamic anomalies, called the *Jagla fluid*.

Zanotti et al. tested the hypothesis on the two phases of liquid water ("liquid–liquid critical point hypothesis") experimentally. Using the *differential scanning calorimetry* (DSC) and the neutron spectroscopy, the authors observed two states of the supercooled water differing in density and the first-order transition between these states. The existence of this critical point could explain, inter alia, the anomalous increase of the isothermal compressibility and isobaric specific heat on cooling.

Experimental study of the water molecule dynamics in SWCN showed that at 228 K the temperature dependence of the relaxation time is changed from the Vogel–Fulcher–Tammann law

$$\tau = \tau_0 \exp\left(\frac{B}{T - T_0}\right) \qquad ...(6)$$

to the Arrhenius-type dependence

$$\tau = \tau_0 \exp\left(-\frac{E}{\kappa T}\right),$$

where E is the activation energy.

The authors assume that this moment corresponds to the first order transition from the low density ($\rho < 1.02$ g cm^{-3}) liquid water to the high density ($\rho > 1.14$g cm^{-3}) liquid water. Similar results on the change of the temperature dependence of the relaxation time were obtained for the water in the pores of 50 nm diameter in the ceolites. It should be noted, however, that dynamic water transition observed by E. Mamontov et al. in SWCN with the inner diameter of 14Å was not detected in the double-wall carbon nanotubes with the inner diameter of 16Å.

Since heterogeneous nucleation could be experimentally circumvented, the homogeneous nucleation theory gives an upper bound to the maximum supercooling. A homogeneous nucleation of the pure water is observed at –39 °C; more deep supercooling can be achieved underthe high pressure (up to –92 °C) and in nanopores (it should be noted that, on the other hand, the glassy state of the water in the pores and at the silica surface of Vycor could exist at the room temperature); the water dynamics corresponding to a liquid state is observed at low temperatures in the carbon nanotubes. When the temperature is lowered further, tube-like solid structures are observed named by the author's ice-nanotubes (ice-NTs) or a square-ice sheet wrapped into a cylinder inside the carbon nanotube.

The water is not unique in rather high degree of supercooling that can be achieved: the ratio of the homogeneous nucleation temperature to the freezing temperature at normal conditions is about 0.85 for the water while this ratio is 0.79 for phosphorus and 0.75 for ammonia.

Note that the value of –39 °C as the homogeneous nucleation limit below which even very small droplets would crystallize is in some contradiction with experiments by L.S. Bartell and J. Huang, who observed freezing of droplets formed by condensation of water vapour in the supersonic Laval nozzle to cubic ice I_c at temperatures close to 200 K.

Recently the possibility of the existence of ice at room temperature has been established. Computations performed by A. Wissner-Groos and E. Kaxiras showed that a layer of diamond coated with sodium atoms will keep a thin layer of water in the frozen state up to 108 degrees of Fahrenheit (42 °C). This discovery (if confirmed by experiments) will greatly advance biocensors and other implanted devices, since an ice layer covering the nonorganic surface will smooth the diamond surface and prevent clotting proteins from attaching the surface.

The electric field of strength about 10^6 Vm^{-1} raises the crystallization temperature of the water in the nanoscale slit to room temperature. Both the restrictions on the water molecule's mobility in the normal to the wall direction and ordering of the dipoles in the electric field help the formation of the stable net of hydrogen bonds and of the critical size embryo. This effect is reversible - decreasing the field strength results in the ice melting.

S. Wei et al. studied experimentally the effects of dipole polarization of water molecules on ice formation under an electrostatic field with strength E in the range 10^3–10^5 Vm^{-1}. When a water molecule is in the external electrostatic field, there is force moment M acting on the molecule given as $M = \mu_0 E \sin \theta$, where μ_0 is the electric dipole moment of the water molecule and θ is the angle between the dipole moment direction and the field direction. The water molecules, excluding the case when the dipole moment is parallel to the field direction, tend to turn in the direction of field to reach the stable state. The potential energy of the interaction U is $\mu_0 E \cos \theta$. Assuming the Boltzmann distribution law, the distribution of water molecules could be written as

$$f(U) = A\exp\left(-\frac{U}{\kappa T}\right) = A\exp\left(-\frac{\mu_0 E \cos\theta}{\kappa T}\right).$$

Evidently, the Boltzmann distribution function attains maximum when all water molecules have their dipoles oriented along the electrostatic field direction that assists in forming the critical nuclei. The authors, however, found that in contrast to the nucleation, the growth time was unaffected by the presense of the field. The effect of the electric field on the nucleation of the ice crystals is known for over one and a half centuries. The electric field could be used to assist the formation of the ice nuclei in the water solution at the given temperature to control the size of grains in the growing polycrystal or the liophyllization time.

Liquids can exist in metastable supercooled state; in contrast, solids cannot be superheated since in contrast to freezing there is no energy barrier for melting. The temperature of solids is always below their (pressure-dependent) bulk melting temperature T_m. When solid temperature is close (from below) to T_m, the phenomenon of premelting occurs: mobile liquid is observed on at one facet of a wide class of materials that includes water, solid rare gases, semiconductors, metals. The premelting is subdivided into the surface melting (at the solid-vapour interface) and the interfacial melting (at the interface between the solid and the foreign substrate). The existence of the interfacial premelting depends on a competition between attraction of the liquid to the solid (called *adhesion*) and attraction of the liquid to itself (*cohesion*). The film thickness d is proportional to the deviation of the temperature from its bulk melting value:

$$d = \lambda(T_m - T)^{-\frac{1}{\nu}},$$

where the parameter ν characterizes the fall-off rate of the inter-molecular forces and equals $\nu = 2$ for the electrostatic and $\nu = 3$ for the van der Waals forces. It could be seen that the film thickness increases rapidly as the temperature tends to the bulk melting temperature, that is, the crystal melts from its surface inwards.

Ice exhibits all aspects of premelting, including grain-boundary melting at junctions between the individual crystals in the polycrystal. Polycrystalline ice contains liquid water due to the impurity and curvature depression of the freezing point in microscopic channels (a few tens of micrometer wide) called veins formed by the tri-grain junctions. Water is also present at nodes separating four grains. Water forms a network distributed thoughout the whole polycrystal.

The so-called "*growth-melt asymmetry*" was observed for pure ice crystals at high pressure (2,000 bar). As the pressure increases, the melting temperature decreases and the system passes through a roughening transition with change of the cylindrical shape into the hexagonal prism due to the formation of the facets. Both growth and melt shapes are hexagonal, but the corners appear to be rotated by 30°. This behavior is explained by the more fast melting of the initial corners where the molecules are weakly bound; the rate of the corners' recession is considerably larger than that of the facets.

Some of the water anomalies are also exhibited by other tetrahedrally coordinated or simply tetrahedral liquids with different bonding varying from the purely ionic nature in the case of BeF_2 to the purely covalent in the case of SiO_2.

Biological Water

Second, the behavior of the water molecules near the hydrophobic or hydrofillic surfaces cardinally differs from their behavior in the bulk. The terms *bond*, *ordered*, *biological* water are used, and the cryobiologists also speak about *unfreezable* water (it is more correct, however, to use the word *unfrozen*, allowing for a nonequilibrium state). It is possible to distinguish *buried* or *inner* water that is held by the hydrogen bonds in the inner hydrophobic regions of the biological molecules and sometimes form clusters. Such water is present in most globular proteins; in spite of the permanent exchange with the exterior molecules with the characteristic times from 10 ns to 1 ms, it should be considered as an intrinsic element of the protein. The term *vicinal* water (from the word vicinity) is usually used in respect to inorganic surfaces.

The importance of the water for the stability and function of biological molecules hardly can be overestimated – the water itself could be considered as the biomolecule and even is called the 21st amino acid. In some cases the normal behavior of the protein molecule is possible only if the water on its surface forms (in terms of the percolation theory) an infinite cluster. The hydrogen bonds from the water to the peptide groups are only slightly stronger than the hydrogen bond between the water molecules. MD computations reproduce the anomalous behavior of the water near the surface.

To establish the hydrogen bond with the protein surface water either donates its proton to protein (to carbonyl/carboxyl and hydroxyl oxygen atoms) or accepts proton from the protein (from amine, amide, and hydroxyl groups that are present in the protein backbone (NH) and in some side chains (NH, NH_2, NH_3)). Water molecule also could mediate interaction between two protein atoms (*"water bridge"*). Hydrogen bond contribution to the solvation energy was determined recently by M. Petukhov et al. for several atom types (NH/NH_2-uncharged groups, H in hydroxyl groups, NH_2/NH_3-charged groups, O-uncharged carbonyl/hydroxyl groups, and O-charged carboxyl groups).

The change of the average orientation of the water molecules near the interface could be assessed by the normalized probability distribution for the angle between the water dipole and the normal to the interface and also for the angle between the normal and the vector joining the hydrogen atoms. In addition to the water/solid interfaces of different nature, the water/carbon tetrachloride interface was extensively studied, since this system serves as a prototype for hydrophobic interactions and in many respects mimics the biological membrane.

The variation of the water properties near the biological surface is caused by the regrouping of hydrogen bonds – by ordering of the water molecules - determined by both the nature of the surface (polar or non-polar) and the spatial restrictions on the structure of the hydrogen bonds net (cutoff of the linear correlation scale); the mobility of the water molecules and the water dielectric constant considerably decrease, the water viscosity increases. The effect of the wall proximity on the crystallization temperature is known from observations on the soil freezing and the water behavior in the pores of minerals. Zangi simulated using MD method the water crystallization in the narrow (several diameters of the water molecule) gap for the case of the inert surface that does not form the hydrogen bond with the water molecule. P. Kumar et al. studied the effect of the interaction potential for the gap containing from two to three layers of the water molecules. Two different interaction potentials were considered: Lennard–Jones and purely repulsive potential. The increase of the pressure shifts the water molecules towards the hydrophobic surfaces and increases the value of the tetrahedral parameter; no pressure effect is observed for the case of hydrophilic surface.

MD computations using the density functional theory show the existence of the thin water layer of the higher density near the hydrophilic surface and the significant increase of the strength of bonds between the water molecules in the first hydration shell near the nonpolar surface in comparison with the molecules in the neighbor shells. The authors studied the ordering of the water molecules in the hydration shells of benzene and cyclohexane. They stress that correct classification of surrounding water molecules into discrete shells could be done either by using a distance cut-off, measuring the distance between the water oxygen to the nearest solute carbon atom, or by Voronoi tessellation of the simulation domain, the latter being more robust. Atoms in contact could be explicitly defined as those sharing a face of their surrounding polyhedra accounting for that fact that being closer than some specified distance is not sufficient to be in contact.

Zewall investigated the evolution of the water structure on the model amphiphillic surface (the specially processed silicon surface was used) exploiting the superfast electron crystallography and found an order of magnitude difference between the characteristic times of the decay of the hydrogen bonds structure for the surface layer and in the bulk. On the other hand, the measurements of the rotational mobility of the water molecules in the protein hydration shell and in the bulk using the method of the magnetic relaxation dispersion give only twofold reduction.

Anomalous water properties also manifest themself in the behavior of the fluid viscosity in the layer confined between two close planar surfaces. If the gap thickness is greater than 8–10 molecular diameters, no effect is observed either for the water or for the standard organic liquids. For the gap between five and eight molecular diameters, however, the effective viscosity increases greatly in the case of normal liquids with much smaller relative change for the water.

Computer simulations show that the changes in the dynamic behavior of the water near the surface depend on the morphology of the latter: while the water dynamics typically slow down near the nonattractive *rough* surface, the dynamics speed up near the nonattractive smooth surface.

The direct proofs of the anomalous dynamic properties of the *biological* water were obtained in the experimental study of the water near the lactose molecule using the absorption spectroscopy in the terahertz range (with the time resolution about a fraction of picosecond) and the water at the protein surface with the femtosecond time resolution; several layers of the water ordered due to the interaction with the heads of lipid molecules are also observed near the surface of the biological membrane. The water molecules adjacent to the lipid bilayer interact strongly with the headgroups of the lipid molecules and this interaction can extend to several molecular diameters from the membrane surface: for the case of 1,2-dipalmitol-*sn*-glycero-3-phosphoholin (DPPC) phospholipids computations based on the local density as a function of the distance from the surface show the existence of four distinct regions of water, the first two corresponding to the first and second solvation shells surrounding the phosphocholine groups. It is anticipated that these water layers can effectively influence the membrane permeability, posing an obstacle to the transport of permeant molecules to the membrane and even be a rate-limiting stage in permeation. The study of the water dynamics in the micelles showed that there is no considerable variation of the water mobility in the core while a significant reduction is registered in the Stern layer that contains the polar groups of the amphiphilic molecules and the counter-ions.

Both direct measurements by the terahertz absorption spectroscopy and MD computations of water near the surface of solute (lactose molecule) show over a hundred water molecules beyond the first solvation shell are affected by the molecule, that is, belong to the hydration layer that extends to approximately 5–6Å from the surface (123 such molecules were found in the experiment, 112 were from computations). Molecules in this hydration layer show slower dynamics in comparison with molecules in bulk. The authors assume that this retardation of water dynamics by solute is the source of the stabilization effect on the proteins and membranes in a dehydrated or a frozen state by solvated mono- and disaccharides.

On the basis of the experimental data and MD computations it was suggested to distinguish three kinds of water molecules - free and bound viz.:

1. Molecules bonded to other water molecules only
2. Molecules having a single bond with the macromolecule's polar group
3. Molecules having two bonds with the macromolecule's polar group

Somewhat more formal classification of hydration water molecules was suggested by M. Nakasako in relation to the water present in protein crystals. The four classes are "inside" (water molecules that occupy cavities in the protein molecules), "contact" (molecules that are located outside the SAS of protein and mediate interactions between adjoining molecules in the crystal; these contact-class hydration sites are formed during the crystallization process), "first-layer" and "second-layer." The electron density in the first-layer hydration shell is greater than that in bulk solvent, thus dielectric properties of hydration water molecules of this class must differ from those in bulk solvent. Hydration water molecules of the first-layer and second-layer classes form aggregates using hydrogen bonds that link together and are indirectly connected to the polar protein atoms forming a network of hydrogen bonds on the protein surface.

The reduction of the water mobility near the solid surface is qualitatively similar to the behavior of the supercooled water; sometimes even an "*equivalent temperature shift*" is determined: for example, the value of this parameter for the water in the nanogap between two smooth *hydrophobic* surfaces is estimated from MD computations as 40K, while M.C. Bellissent-Funel reported the estimated temperature shift of 20 °C that is attributed to the difference life times of the hydrogen bonds. On the other hand, simulation for the case of the *hydrophilic* surface of hydroxide $Mg(OH)_2$ shows that the confinement effect is more akin to the pressure increase. When the temperature decreases, similar to the aforementioned experiments with SWCT, the change of the relaxation time dependence on the temperature is observed approximately in the same temperature range (222 K and 220 K for the water on the surface of DNA and lysozime, respectively).

The discrepancies in the dynamical data of the biological water obtained in the different measurements are related to the difficulties of separating the water and the macromolecule motion in some experimental approaches; for example, when quasi-elastic and inelestic neutron scattering is used, the elastic contribution from the biological molecule itself could be significant. B. Halle noted that reduction of the mobility of the macromolecules in the solution in comparison with the mobility of the equivalent (hydrodynamically) body (a sphere, an ellipsoid) is at least partly explained by the significant surface irregularity.

Crystallization in Heterogeneous Media

Finally, the biological media has a complex microstructure. In the absence of a seed surface, ice formation in the supercooled solution is initiated by the homogeneous or the heterogeneous nucleation. The heterogeneous nucleation dominates if the following nonequality ("*wetting criterium*") is valid:

$$\sigma^{vs} - \sigma^{vl} < \sigma^{ls},$$

where σ^{ij} is the surface tension on the interface between phases i and j, subscripts v, l, and s denote vapour, liquid, and solid, respectively. The nucleation is observed at the cell membrane or at the objects contained in cytoplasm; the importance of the nucleation on the subcellular objects increases as the temperature decreases; heterogeneous nucleation has not been observed in the cells of some hardwood plants and co-existence of ice and supercooled water is possible.

The phase transition in the biological media occurs in the finite temperature interval (in this respect biological tissues are not unique – the same behavior is observed, for example, in freezing of the soils). There are two reasons for the finite temperature interval of the crystallization in the biological media: continuous variation of the solute concentrations in the different tissue compartments of finite volume and the presence of the extracellular matrix.

The freezing process of the pure water and aqueous solutions, being cooled externally, is well known. First, the temperature decreases with time until the maximal supercooling ΔT_m is achieved. Once nucleation occurs, the rapid crystal growth results in the releasing of the latent heat of fusion faster than it can be removed from the system and rise of the temperature to the equilibrium melting temperature. The system temperature then remains constant for some time until all water is transformed into the ice (the so-called "*latent heat plateau*"), then (if external cooling is continued) again decreases.

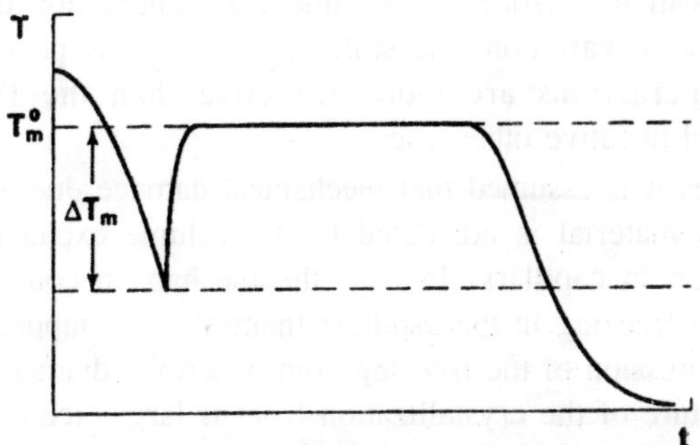

Fig. 5.1. Freezing of pure water.

The behavior of the freezing aqueous solution displays a few differences:

1. The absolute temperature of supercooling, i.e. the nucleation temperature, is usually higher than that for the pure water, since the presence of solutes promote nucleation
2. The temperature rise after nucleation is smaller, since the equilibrium melting temperature of the solution is depressed
3. The temperature on the "latent heat plateau" is not constant, but slowly decreases due to the continuous growth of the solute concentration and, hence, continuous lowering of the equilibrium melting point of the solution
4. There is another peak on the temperature vs. time curve related to the heat release at the point of the eutectic crystallization; cooling below T_{Eu} will be continued only after the solution has solidified completely.

The behavior of water in porous media, including freezing, is known to differ from the behavior of bulk water. The relative strength

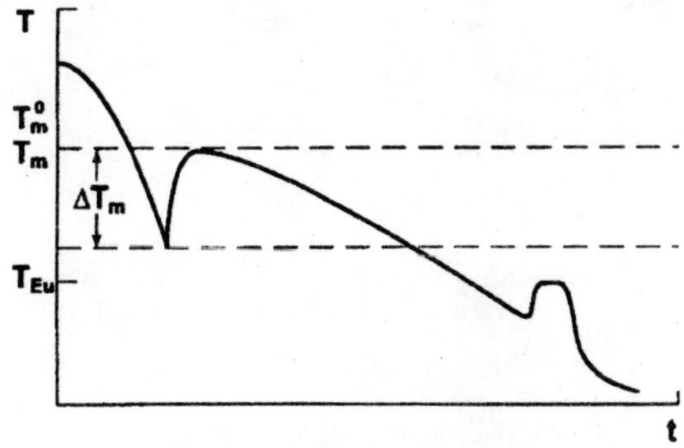

Fig. 5.2. Freezing of solution.

of the fluid–wall interaction to the fluid–fluid interaction determines the freezing point variation: the shift $T_{f,pore} - T_{f,bulk}$ is positive if the fluid–wall interactions are more attractive than the fluid–fluid interaction and negative otherwise.

Sometimes it is assumed that mechanical damage due to freezing in the porous material is attributed to the volume expansion at the phase transition in capillars. In fact, the mechanism could be more involved since freezing in the capillars themselves is suppressed by a significant depression of the freezing point when the diameter is small and the curvature of the crystallization front is large. Ice crystals are first formed in the relatively large voids and are called ice lenses. These crystals are not insulated and the water is supplied through the capillars to support further ice growth. Finally, large enough crystals act mechanically on the porous matrix causing its deformation. In the case of wet soil freezing, this phenomenon is known as frost heave. It alters landscapes, destroys roads, runways, pipelines, building foundations in countries with severe winters, and redistribute methane in the permafrost regions.

6

Imitation Models

Imitation models are aimed at the simulation of functioning of large systems. Usually a model includes a description of the possible system states along with some elementary processes that perform a transition of the system from one state to another. A model conserves both the logical structure of such discrete events as well as their casual relationships and, thus, the time sequence of events.

The imitation model usually does not try to reproduce the real transition rates and thus its predictions are qualitative only. However, such models allow one to gain a general understanding of system working, interaction of its components, and of the relative importance of different parameters that govern the behavior of the system. Imitation models, as a rule, are rather fast; frequently, an analytical solution is available for some limiting cases.

Intracellular Ice Formation

Zhang et al. considered a model of the tightly packed cells that is composed of several concentric ring layers of small malignant cells surrounded by the extracellular shell. The layers are numbered from outside inwards. The authors stated that four cell layers correspond to the pattern observed experimentally. Because of the smallness of this structures (radius is about 80 μm), the temperature was assumed to be spatially uniform, while its variation in time was determined by the solution of the one-dimensional Pennes equation. The evolution of the volume of the first (outer) cell layer is determined by the water transport:

$$\frac{dV_1}{dr} = -\frac{L_p A_1 R_g T}{\upsilon_w}\left[\ln x_1 - \frac{L\upsilon_w \rho}{R}\left(\frac{1}{T_R} - \frac{1}{T}\right)\right] - \frac{L_{p,c} A_2 RT}{\upsilon_w}\ln\frac{x_1}{x_2},$$

where V_1 is the first layer volume, A_1 and A_2 are the permeation surfaces associated with the first and second layers, υ_w and ρ are the water molar volume and density, L is the latent heat, T_R is the reference temperature, R_g is the universal gas constant. It is assumed that for the cells in close contact the hydraulic permeability $L_{p,c}$ = 0.5 L_p, x_i is the mole fraction of the intracellular water in the *i*th cell layer:

$$x_i = \frac{V_{0,i} - V_{b,i}}{(V_{0,i} - V_{b,i}) + \gamma_s \eta_{s,i} \upsilon_w},$$

where V_0 is the initial cell volume, V_b is the osmotically inactive cell volume, γ_s is the dissociation constant for salt, $\eta_s = C(V_{0,i} + V_{b,i})$ is the mole fraction of the salt ions, and C is the cell osmolarity.

The water transport for other cell layers is driven by the concentration differences, therefore

$$\frac{dV_i}{dt} = \frac{L_{p,c} A_i RT}{\upsilon_w} \ln \frac{x_i}{x_{i-1}} - \frac{L_{p,c} A_{i+1} RT}{\upsilon_w} \ln \frac{x_i}{x_{i+1}}$$

for the intermediate layers (i = 2,…, n – 1) and

$$\frac{dV_n}{dt} = -\frac{L_{p,c} A_n RT}{\upsilon_w} \ln \frac{x_n}{x_{n-1}}$$

for the central cell region. The authors used the model by Toner and Cravalho to write the probability of the intracellular ice formation for the innermost (central) cell layer $i = n$ as P_{IIF} = 1 – exp (–τ), where

$$\tau = \int_{T_{seed}}^{T} A\Omega_0 \left(\frac{T}{T_f}\right)^{1/2} \frac{\eta}{\eta_0} \frac{A}{A_0} \exp\left(\frac{-\kappa_0 \left(T_f / T_{f0}\right)^{1/4}}{(T - T_f)^2 T^3}\right) dt,$$

where Ω_0 and κ_0 are the cell-type-dependent thermodynamic and kinetic parameters of homogeneous nucleation, A is the surface area, T_f is the freezing temperature of the intracellular fluid whose dependence on the solute mole fraction could be expressed as

$$T_f = \frac{T_R L}{L + T_R((1 - x_1) - (1 - x_{i,0}))}$$

and η is the viscosity depending on both temperature and solute concentration as

$$\eta = \eta(T, \phi_m) = 0.139 \times 10^{-3} \left[\frac{T}{225} - 1\right]^{-1.664} \exp\left(\frac{2.5\phi_m}{1 - 0.609375\phi_m}\right),$$

the volume fraction of solute ϕ_m could be approximated by $V_{b,i}/V_i$ and the subscript "0" denotes the isotonic state.

Zhang et al. assumed the propagation of the intracellular ice in the densily packed cluster of four concentric spherical layers to be instantaneous. Then for the probability of IIF for the *i*th cell layer the authors write

$$P_{\text{IIF}}(i,t) = (1 - \exp(-2\tau))\, P_{\text{IIF}}(i-1,t) + (1 - \exp(-\tau))(1 - P_{\text{IIF}}(i-1,t)).$$

The heat transfer problem was considered under the following assumptions:

1. Phase transition occurs between 0 and −10 °C
2. The density of the media does not change upon the phase transition
3. The thermal conductivity is constant, but different in the frozen and unfrozen regions
4. Blood perfusion rate and metabolic heat generation reduce linearly with temperature down to zero at the crystallization front
5. The specific heat is defined as an effective property incorporating the latent heat
6. The thermophysical properties of the healthy tissues and those of the tumor are the same

Computations showed that only cells in the first layer suffer considerable dehydration while for the cells in the inner layers IIF is more important. In accordance with observations of Hong and Rubinsly, IIF occurs at lower cooling rates than are typical for the cells in suspension.

Later Karlsson showed, however, that the assumption about the infinite rate of the intracellular ice propagation is not correct and the finite speed of the intracellular ice propagation should be taken into account.

Even if the propagation is instantaneous, the probability of IIF depending on the layer number does not make sense. As Karlsson righteously noted, under this assumption intracellular ice will appear in all cells simultaneously upon the first nucleation event in any cell belonging to the cluster, and the total nucleation rate in some cell in the cluster $J(t)$ will be given by the sum of the IIF probabilities in the individual cells

$$J(t) = \sum_{j=1}^{n} \sum_{i=1}^{m_j} J_{i,j}(t),$$

where it has been assumed that the number of cells in the *j*th layer is m_j. If the nucleation could be described as a Poisson process with an average rate $J(t)$, the cumulative probability of ice nucleation in the cluster will be expressed as

$$P_{\text{cluster}}^{\text{IIF}} = 1 - \exp\left[-\int_0^t J(t)dt\right].$$

Irimia and Karlsson considered a pair of contacting cells as a system with a finite number of states, namely, three-state system:

(0) There is no ice in the either of cells

(1) There is ice in one of the cells

(2) There is ice in both cells

The state of the ensemble of cells could be described by two parameters, for example, by the probability of "unfrozen" $P_0 = N_0/N$ and "singlet" $P_1 = N_1/N$ states (clearly, the probability of the "duplet" state $P_2 = 1 - P_1 - P_2$). An evolution of the cell ensemble is governed by two independent stochastic processes – a spontaneous nucleation of the intracellular ice J_s and the ice propagation through the intercellular contact J_p. Evidently,

$$\frac{dN_0}{dt} = -2J_s N_0 \qquad \text{...(1)}$$

and

$$\frac{dN_2}{dt} = (J_s + J_p)N_1. \qquad \text{...(2)}$$

Introducing a dimensionless time

$$\tau = \int_0^t J_s dt$$

and a dimensionless rate of ice propagation

$$\alpha = \frac{J_p}{J_s},$$

(1) and (1) could be written as

$$\frac{d}{d\tau}\begin{bmatrix} P_0 \\ P_1 \end{bmatrix} = \begin{bmatrix} -2, & 0 \\ 2, & -(1+\alpha) \end{bmatrix}\begin{bmatrix} P_0 \\ P_1 \end{bmatrix}.$$

It can easily be seen that

$$P_0(\tau) = \exp(-2\tau).$$

In the case of constant α an analytical solution for P_1 is written as

$$P_1(\tau) = \begin{cases} \frac{2}{1-\alpha}(\exp(-(1+\alpha)\tau - \exp(-2\tau)), \alpha \neq 1 \\ 2\tau \exp(-2\tau) \qquad \alpha = 1 \end{cases}$$

Maximal probability of the singlet state

$$P_1^* = \left(\frac{1+\alpha}{2}\right)^{(1+\alpha)/(1-\alpha)}$$

is achieved at

$$\tau^* = \frac{1}{1-\alpha}\ln\left[\frac{2}{1+\alpha}\right].$$

Ice formation and propagation in the aggregates of cells was considered under assumption that the rate of nucleation in the current cell is defined as

$$J = J_s + mJ_p, \quad \text{...(3)}$$

where m is the number of the neighbor cells. The total number of states M in the system containing N cells comprising all distinct combinations of states (either frozen or unfrozen) of the individual cells is

$$M = 2^{N-1} + 2^{(N-1)/2}.$$

In the ensemble of the identical cell aggregates, an evolution of the vector P of the probabilities P_ξ of each state ($\xi = 0, \ldots, M-1$) is described as

$$\frac{dP}{d\tau} = Q(\alpha)P,$$

where the matrix Q contains all the rates of transitions between different states. This equation is sometimes referred to as Kolmogorov differential equation.

An analytical solution for the case $N = 4$ was presented in the cited paper. The formation of the intracellular ice for large N was simulated using Monte-Carlo method. The tissue sample is described by a regular lattice in which each lattice site represent a cell that could be in one of the two states (frozen/unfrozen). Ice propagation occurs between immediate neighbors. Assuming that ice propagation could be described by a Poisson process with a rate given by (3), Irimia and Karlsson wrote out the probability that an unfrozen cell j will freeze within a time interval Δt as

$$P_j^{\text{IIF}}(\Delta t) = 1 - \exp\left[-\int_t^{t=\Delta t}\left[J_j + m_j(t')J_p(t')\right]dt'\right],$$

where m_j is the number of nearest neighbors of the cell j that are frozen.

If one further assumes that both α and m_j are constant during this time interval Δt, one get a *homogeneous* Poisson process

$$P_j^{\text{IIF}}(\Delta t) = 1 - e^{-(1+m_j\alpha)\Delta t}.$$

To enforce the validity of the assumption on the constancy of m_j during the time interval Δt, the authors required the probability of IIF occurring anywhere in the sample during this time interval to be bounded from above by some small probability ε, leading to the constraint on the time step

$$\Delta t \le \left\{ \sum_j \frac{-\ln(1-\varepsilon)}{(1+m_j\alpha)}, \right.$$

where the sum is taken over all unfrozen cells in the sample.

For time intervals $\Delta\tau$ satisfying this equations, due to the smallness of the probability of simultaneous spontaneous nucleation (s) and ice propagation from the neighbor cell (p), these mechanisms could be considered as independent to yield

$$P_j^s(\Delta\tau) = 1 - e^{-\Delta\tau} \approx \Delta\tau, \qquad \text{...(4)}$$

$$P_j^p(\Delta\tau) = 1 - e^{-m_j\alpha\Delta\tau} \approx m_j\alpha\Delta\tau, \qquad \text{....(5)}$$

respectively, and the total probability of IIF in the cell j is approximately

$$P_j^{\text{IIF}}(\Delta\tau) \approx P_j^s + P_j^p \approx (1 + m_j\alpha)\Delta\tau.$$

The authors thus choose the time step for the Monte Carlo computations as

$$\Delta\tau = \frac{\varepsilon}{(1 + n\alpha)N},$$

where the maximum number of the nearest neighbors m is $m = 2$ for the linear cell arrays and $m = 4$ or $m = 6$ for the two-dimensional cell aggregates, depending on the cell shape and packing. In computations either one or two independent quasi-random numbers were used, depending on whether the distinction between different nucleation mechanisms is to be monitored.

One-dimensional array of 100 cells has been considered by Irimia and Karlsson while Stott and Sumper solved two-dimensional problems for the number of neighbors $m = 6$ and $m = 4$.

When the domain contain both pathological ("t") and healthy ("h") tissues, the nucleation rate is written as

$$J = J_s + m^t J_p^t + m^h J_p^h.$$

A new parameter that characterize the ratio of the ice propagation speeds in two kinds of tissue arises $\beta = J_p^t / J_p^h$. Computations for a wide range of the ice propagation speed in the healthy tissue revealed three growth regimes differing by the dependence of the pathological tissue destruction on the parameter β.

Large scale Monte Carlo computations were performed by Stott et al., accounting for the characteristic length scale of the thermal gradient in the vicinity of the cryoprobe for a two-dimensional tissue sample containing 10^4 cells.

A sample was represented by a lattice of 99 × 99 cells with each interior cell interacting with four neighbors. The computational domain was divided into two regions corresponding to the tumor and healthy tissue. The centrally located tumor occupied one half of the domain. Similarly to the previous analysis, dimensionless rates of ice propagation were introduced:

$$\alpha^h = \frac{J_p^h}{J_s^h}, \quad \alpha^t = \frac{J_p^t}{J_s^h}$$

and

$$\gamma = \frac{J_s^t}{J_s^h},$$

the latter being the dimensionless rate of spontaneous II nucleation. The use of the same denominator in the definition of αs allowed one to define the relative amplitude of the ice propagation rates in the healthy and tumor tissues β, which can also be written in terms of dimensionless propagation rates α^h and α^t as $\beta = \alpha^t/\alpha^h$.

For the heterotypic cell contact the author used the slowest propagation rate $J_p^{jk} = \min(J_p^h, J_p^t)$.

To imitate the effect of the spatial temperature nonuniformity around the cryoprobe, the authors have assumed that both the spontaneous nucleation rates and the ice propagation rates decay exponentially with the distance from the cryoprobe, that is,

$$J_s^{jk} = J_s^h e^{-\frac{r_{jk}}{\lambda}} \quad \text{and} \quad J_s^{jk} = J_s^t e^{-\frac{r_{jk}}{\lambda}}$$

for the cell jk in the healthy and tumor regions, respectively, and similarly

$$J_p^{jk} = J_p^h e^{-\frac{r_{jk}}{\lambda}} \quad \text{and} \quad J_p^{jk} = J_p^t e^{-\frac{r_{jk}}{\lambda}}$$

for the homotypic interactions within healthy or tumor regions and

$$J_p^{jk} = \min\left(J_p^h, J_p^t\right) e^{-\frac{r_{jk}}{\lambda}}$$

for the boundary between the tumor and healthy tissues. Here the dimensionless distance of the cell jk from the surface of the cryoprobe r_{jk} is related to the effective tumor radius R

$$r_{jk} \equiv \frac{\sqrt{x_{jk}^2 + y_{jk}^2}}{R},$$

and λ is a parameter that specifies the rate of decay of the nucleation and propagation rates away from the cryoprobe (to the e^{-1} at the distance λ from the probe).

The above described Monte Carlo algorithm is based on the discrete nucleation events. It is possible to simulate "continuous" nucleation exploiting the approach developed by Gillespie for the general chemical kinetics. To accomplish it, a joint probability function on the discrete variable j specifying the cell where the next IIF event will occur and on the continuous variable $\delta\tau$, representing the time interval until the occurrence of this IIF. To generate a pair of these random variables, two random numbers r_1 and r_2 uniformly distributed over the unit interval should be draught to permit the estimation of $\delta\tau$ as

$$\delta\tau = \frac{1}{a_0}\ln\left(\frac{1}{r_1}\right),$$

where a_0 is the specified parameter of the probability density function – the overall rate of the intracellular ice formation in the tissue – while the cell number j where the IIF will take place is determined by the following inequalities:

$$\frac{1}{a_0}\sum_{m=1}^{j-1} h_m c_m < r_2 < \frac{1}{a_0}\sum_{m=1}^{j} h_m c_m,$$

where h_m is the state parameter in the Gillespie algorithm and c_m is the rate of conversion of the mth cell from the unfrozen to frozen state.

Sumpter notes that typical size of the cluster that is practically reasonable to be simulated by the described Monte Carlo algorithm is about 10^5 cells that correspond to the tissue size 1–10 mm^2 in the biological sample and suggested a modification of the above mentioned continuous Johnson–Mel–Avrami–Kolmogorov model that describes the kinetic of phase transformation by evolution of the transformed volume fraction $X(t)$ in time as

$$X(t) = 1 - \exp(-kt^n)$$

where k and n depend on the geometric factors and on the mechanisms of phase transition.

Cellular Automata and Related Methods in Cryobiology

Cellular automata and other methods based on the finite number of the system states are rarely explicitly referred to as imitation models;

it is also claimed that these methods provide the quantitatively correct simulation. Still, the word "imitation" is adequately describes both the underlying structure of these methods (a finite number of states associated with sites on some spatial lattice) and the computational procedure (a set of rules that define transition to the new state depending on the state of the current site and its neighbors). The difference in the scale (level) only: while classical imitation models deal with large components of the complex system (macrolevel), the cellular automata and related simulation methods exploit the same strategy on the microlevel. Using popular in simulation studies terminology, one can say that classical imitation models use coarse-grained system decomposition and cellular automata - fine-grained.

The cellular automata method originates from the well-known Ising (two- state) model in physics for the evolution of the magnetized systems and its generalization - the Potts (multi-state) model. These methods are easily parallelized and that is one of the reasons of their growing popularity. The most advanced implementation of the cellular automata approach is probably lattice Boltzmann method (LBM), also referred to as lattice Boltzmann machine. In some sense LBM is the minimal realization of the Boltzmann kinetic equation, thus macroscopic quantities of interest are computed by averaging the lattice representation of "*distribution function.*" LBM was applied to a few cryobiology problems including the ice recrystallization and for the solution of the bioheat Pennes equation.

7

MICROSCOPIC MODELS

Contrary to the well-known witticism by the English statesman (sometimes wrongly attributed to M. Twain), statistical methods are the powerful tool for the study of the phenomena characterized by the intrinsic stochasticity that are very difficult if not impossible to describe by the pure analytical methods. These approaches are divided into two groups - Monte Carlo (MC) and Molecular Dynamics (MD) simulations. With still ongoing advances in computer hardware (even if some deviations from the Moore law on the constant rate of gaining the computing speed are noticed) it is expected that their importance will increase in future, allowing simulation of the more complex systems for the longer physical time.

When the subject of the study is phase transition, MC or MD approach could be used to generate the relation between the order parameters and the free energy of the system via the bottom-up approach. If N order parameters $\varphi_1, \varphi_2, \ldots, \varphi_N$ are specified, the free energy density $\Lambda[\varphi_1, \varphi_2, \ldots, \varphi_N]$, also called the potential of mean force or Landau free energy, is related to the microscopic Hamiltonian H as

$$\Lambda[\varphi_1, \varphi_2, \ldots, \varphi_N] = \int dr_1 dr_2 \cdots dr_N \exp(-\beta H(r_1, r_2, \ldots, r_N))$$
$$\times \delta(\phi_1 - \phi_1')\delta(\phi_2 - \phi_2') \cdots \delta(\phi_N - \phi_N').$$

Here δ is the Dirac delta function and the free energy F is expressed as

$$\exp(-\beta F) = \int d\phi_1 d\phi_2 \cdots d\phi_N \exp(-\beta\Lambda[\varphi_1, \varphi_2, \ldots, \varphi_N]).$$

In the course of the simulation, the free energy is computed by collecting the histograms of the distribution of the order parameters.

For the sufficiently long simulation these histograms are proportional to the joint probability of the order parameters $P[\varphi_1, \varphi_2, \ldots, \varphi_N]$, which is related to the free energy density as

$$\beta\Lambda[\varphi_1,\varphi_2,\ldots,\varphi_N] = -\ln(P[\varphi_1,\varphi_2,\ldots,\varphi_N]) + C,$$

where C is a constant.

The growth rate of the crystal could be related to the increase of the number of "solid" particles N_s as

$$R = \frac{d}{A/a}\frac{dN_s}{dt},$$

where d is the interplane distance parallel to the interface, A is the total area of the interface, and a is the specific area per particle. The problem is no net growth in the equilibrium. Thus, to compute the growth rate from the equilibrium simulations, an Onsager's hypothesis is invoked that state that equilibrium fluctuations decay according to the macroscopic laws.

Monte Carlo Method

Monte Carlo simulation methods are especially useful in studying systems with a large number of coupled degrees of freedom. The major components of MC algorithm are the following ones:

1. Random number generator - a source of random numbers uniformly distributed on the unit interval
2. Sampling rule - a prescription for choosing from the possible states of the model physical system
3. Scoring - accumulation of the computational results to get estimates for the quantities of interest
4. Error estimation - an estimate of the statistical error (variance) as a function of the number of trials and other quantities

The algorithm implementation is easily divided into two largely separated stages - generation of the "raw" data and data analysis.

The MC method based on random sampling is very simple; however, its naive application will be extremely inefficient. One of the most important aspects of the MC simulation is the use of an advanced *sampling technique* that could be rather sophisticated. The idea is that one should not pick configurations randomly, but should draw them according to their Boltzmann weight. Different update rules are usually subdivided into more conceptually simple local and more involved nonlocal algorithms. The important ingredient of MC simulation is the use of various variance reduction techniques to decrease the computational time for Monte Carlo simulation.

Update Algorithms

The most popular is undoubtly the local update algorithm that was suggested over half a century ago by Metropolis and later named after him. The algorithm is formulated as follows. If E and E' are the current energy and the energy after the tentative update, respectively, the probability of the acceptance of this update (the transition probability $W(\sigma_i \rightarrow \sigma_i')$ from the state σ_i to the state σ_i') is given by

$$W(\sigma_i \rightarrow \sigma_i') = \begin{cases} 1, & E' < E \\ \exp[-\beta(E' - E)], & E' \geq E, \end{cases}$$

or, more compactly, as

$$W(\sigma_i \rightarrow \sigma_i') = \min\{1, \exp[-\beta(E' - E)]\}.$$

The key feature of the Metropolis algorithm is the finite probability of the acceptance of the unfavorable updates that increase energy (this probability tends to zero in the case of zero temperature expressed via the parameter β as $\beta \rightarrow \infty$). It is this feature that prevents from getting trapped in the local minimum. This algorithm has a wider range of applicability, in particularity, it is a basis of the simulated annealing - the general optimization method that provides under some rather mild conditions finding the global minimum while avoiding local minima.

There are other local update algorithms such as Heat–Bath or Glauber, for example; the former is usually applied to the lattice models and in its simplest form to the systems with discrete degrees of freedom with a small number of allowed states.

Different patterns could be used to update degrees of freedom, including the random one or the random permutation, or just the lexicographical (sequential) order. This choice has small effect on the qualitative algorithm behavior, but could significantly influence the computational time.

The nonlocal algorithms such as the multigrid schemes or the cluster-update methods improve the performance of computations at the cost of generality. Being more specified than the local update algorithms, they could encounter difficulties in the application to the new models.

Among analysis approaches, different reweighting techniques (single-histogram reweighting, multi-histogram technique) should be mentioned.

One of the most important reasons for the frequently large time needed for the simulation to give the results with the acceptable accuracy is the presence of the so-called "rare events." Thus, two strategies for the computations' acceleration were developed. The first that could be named "rare event avoidance" tries to connect the important parts of the phase space by paths that go around rare-event regions that, hence, could not be studied directly. The second strategy - let us call it "rare event tolerance" - is aimed at enhancing (frequently artificially) the probability of the rare event states.

In the early 1990s, Berg and Neuhaus suggested the multicanonical algorithm (MUCA) for the MC method. The method, in contrast to the conventional MC, is based on non-Boltzmann weight factors, allowing to get a uniform potential energy distribution. The random walks in this energy space require significantly smaller number of simulation steps then conventional canonical MC. MUCA MC splits simulation into two stages: the estimation of the weight factors and the production run itself. However, while the canonical MC gives estimates of the observable physical variables for the fixed temperature and other parameters (extrapolation in the vicinity of this temperature value is possible using reweighting techniques), MUCA MC provides the estimates for the range of temperatures or other parameters. The methods proved to be very efficient for a number of problems, including the phase transition of the first order because it allows to enhance the weights of rare configurations.

Maggs suggested a generalization of the Metropolis update algorithm called by the author "multiscale Monte Carlo," which is based on the collective updates that shift blocks of particles for the distance that is intermediate between the particle size and dimensions of the computational cell and avoids the so-called hydrodynamic slowing down.

The discussion of the intricates of MC methods and algorithmic tricks developed over years to accelerate computations is beyond the present work. The reader is kindly asked to consult numerous books and papers on this subject, for example, the above cited introduction by Janke or the monograph by Robert and Casella.

Molecular Dynamics Methods

Molecular dynamics (MD) per se and as an element of more complex computational methods is one of the main numerical approach to the problems of biophysics; its applications to the cryobiology problems are scarce up to day.

Classical Molecular Dynamics

In the classical MD method, the motion of atoms (or sometimes molecules or other units such as specific residues of proteins) that interact via empirical potential are monitored using the Newtonian equations for the point mass motion. Usually the potential energy of the system could be represented as a sum of several contributions as

$$E(X) = \sum_{\text{bonds } i} S_i^b (b_i - \bar{b}_i) + \sum_{\text{bond angles } i} S_i^\theta (\theta_i - \bar{\theta}_i)^2$$

$$+ \sum_{\text{dihedral angles } i} \sum_n \left(\frac{Vn_i^\tau}{2} [1 \pm \cos(n\tau_i)] \right) + \sum_{\text{atoms } i<j} \left(-\frac{A_{ij}}{r_{ij}^6} + \frac{B_{ij}}{r_{ij}^{12}} \right)$$

$$+ \sum_{\text{atoms } i<j} \left(\frac{Q_i Q_j}{D(r_{ij}) r_{ij}} \right) + \cdots. \qquad \ldots(1)$$

Here the symbols *b*, *θ*, *τ*, and *r* refer to the bond lengths, bond angles, dihedral angles, and interatomic distances, respectively; the bar symbol represent the equilibrium value. Each dihedral angle could be associated with several rotational indices *n* (the third term in the right hand side). The last term corresponds to the Coulomb interaction where *Q* are the atomic partial charges and *D*(*r*) is the spatially dependent dielectric function. *X* represents a collection of the Cartesian coordinates of atoms. Other terms also could be added to account, for example, for hydrogen bonds or planar moieties.

The Newtonian equations of motion that had to be solved at every timestep of MD method are written as

$$M\dot{V}(t) = -\nabla E(X(t)), \qquad \ldots(2)$$

$$\dot{X}(t) = V(t), \qquad \ldots(3)$$

where M is the diagonal mass matrix, dot denotes differentiation with respect to time.

Sometimes frictional and/or random-force terms are added similarly to the phenomenological Langevin equations to yield

$$M\dot{V}(t) = -\nabla E(X(t)) - \gamma MV(t) + R(t), \qquad \ldots(4)$$

$$\dot{X}(t) = V(t). \qquad \ldots(5)$$

Here γ is the damping constant. The random Gaussian force *R*(*t*) should have a zero mean value

$$\langle R(t) \rangle = 0$$

and a specified autocovariance matrix

$$\langle R(t) R(t')^T \rangle = 2\gamma\kappa TM\delta(t - t').$$

Here superscript T denotes a matrix transpose.

The leap-frog integration method introduced by L. Verlet half a century ago for the Newtonian equations of motion is trivially extended to the above modified equations incorporating damping and random force. This method is based on the use of the intermediate time level $t^{n+1/2} = t^n + \Delta t/2$ and can be written as

$$V^{n+\frac{1}{2}} = V^n + M^{-1}\frac{\Delta t}{2}[-\nabla E(X^n) - \gamma M V^n + R^n], \quad \text{...(6)}$$

$$X^{n+1} = X^n + \Delta t V^{n+\frac{1}{2}}, \quad \text{...(7)}$$

$$V^{n+1} = V^{n+\frac{1}{2}} + M^{-1}\frac{\Delta t}{2}[-\nabla E(X^{n+1}) - \gamma M V^{n+1} + R^{n+1}]. \quad \text{...(8)}$$

Ab Initio Molecular Dynamics (AIMD)

Classical MD simulations based on use of the precomputed empirical interaction potentials are not able to describe quantum phenomena that are underlying such processes as proton transfer among water molecules or between the water molecule and a foreign atom or other chemical processes involving bond breaking and forming.

Another deficiency of current force fields is their failure to correctly include polarization effects, although polarizable force fields have been suggested.

The variant of MD as ab initio (AIMD) solves these problems by combining "on the fly" electronic structure calculations with finite temperature dynamics. Naturally, AIMD simulations are substantially more expensive than calculations based on empirical force fields.

The most important element in an AIMD calculation is the representation of the electronic structure. Clearly, calculation of the exact ground-state electronic wavefunction is intractable, and approximations must be used. The electronic structure theory employed should be reasonably accurate yet not too computationally demanding. One formulation of the electronic structure problem that satisfies these criteria is density functional theory (DFT). DFT formulates the many-electron problem in terms of the electron density, rather than the many-body wavefunction. Thus, in principle, the central quantity is a function of just three rather than of 3N variables, a fact that renders calculations based on DFT computationally tractable. The basic problem of DFT is that the energy of a quantum many-body system can be expressed as a unique functional of its density. By minimizing the density functional over all densities that give rise to a particular number of electrons, one obtains the ground state density and energy for a given system. Unfortunately, the explicit and unique form of this functional is not known. However, in the orbital-based formulation of DFT by Kohn and Sham, reasonable approximations to the density functional have been developed. In the Kohn–Sham formulation, the energy is expressed in terms of a set of *n* occupied single-particle orbitals and the *N* nuclear positions.

An extensive comparison of different sampling approaches in the study of the liquid water properties by CarParrinello MD simulations with different values of the fictitious electron mass in the microcanonical and canonical ensembles, Born–Oppenheimer MD in the microcanonical ensemble, and Metropolis Monte Carlo in the canonical ensemble was reported by a group of researches.

Hybrid quantum mechanical/molecular mechanics (QM/MM) approaches are aimed at getting the best of both of these worlds by subdividing the region of interest into inner and outer ones, with solving QM equations in the inner domain only while accounting for the environment (outer) effects at the atomistic level. The first practical application of this approach was to the field that could be named *computational enzymology*.

Acceleration Methods

One is the most common acceleration methods allowing to the increase of the integration time step consists in constraining the fastest degrees of freedom by augmenting the equations of motion via Lagrange multipliers. The iterative procedure of this approach is formulated in the following way:

$$V^{n+\frac{1}{2}} = V^n - M^{-1}\frac{\Delta t}{2}\nabla E(X^n) + G(X^n)^T \lambda^n, \quad \text{...(9)}$$

$$X^{n+1} = X^n + \Delta t V^{n+\frac{1}{2}}, \quad \text{...(10)}$$

$$G(X^{n+1}) = 0, \quad \text{...(11)}$$

$$V^{n+1} = V^{n+\frac{1}{2}} - M^{-1}\frac{\Delta t}{2}\nabla E(X^{n+1}), \quad \text{...(12)}$$

where the vectors λ^n are determined from the equation $G(X^{n+1})$. The matrix G is $m \times m$ Jacobian of the vector g, which specifies the m constraints, and λ is the vector of m constraint multipliers.

One of the critical components that determines the overall computational cost is the treatment of the long-range forces. While the number of bonded forces – presented in the energy expression by the terms involving bond lengths, bond angles, and dihedral angles – grow linearly with the number of atoms, the cost of computing the non-bound – van der Waals and electrostatic – forces depends quadratically on the size of the system. The common cut-off procedures that consist in specifying the distance over which the inter actions are neglected could lead to errors that can not be estimated a priori and are used now for the van der Waals forces only while the electrostatic interactions are treated using such techniques as multipoles or Ewald summations.

Parallel Computations

The computer-intensive character of MD forces one to exploit simplified and hybrid models of the macromolecules and/or to implement simulation codes on the multiprocessor systems. The efficiency of the parallel computations is characterized by the speed-up $S_n = T_1/T_n$ or an efficiency $E_n = T_1/nT_n$ (T_n is the execution time on the system consisting of n processors) that can be written as a product of the three factors: parallel efficiency, numerical efficiency, and uniformity of the load balance. The first one is the ratio of the execution time to the total time that includes computations, data exchange, and waiting for the messages $E_n^{\text{par}} = T_{\text{comp}}/(T_{\text{comp}} + T_{\text{comm}})$ (the time needed to send/receive messaged could be reduced using asynchronous regime – overlapping computations and communications). The numerical efficiency reflects the increase of the volume of computations (e.g., the number of iterations) due to the parallelization of the algorithm and varies broadly depending on the problem and algorithm type. Load balance could be considered statically or dynamically. Ideally,

1. The number of cells or particles for a given processing unit should be proportional to its productivity
2. The volume and frequency of the interprocessor data exchanges should be minimal

One could additionally require the number of the processor pairs in need of data exchange to be minimal. Evidently, some trade-off should be found: minimal volume of communications is reached for a single processor. Thus, one should solve a minimization problem for $T = T_{\text{comp}} + T_{\text{comm}}$ subjected to certain restrictions. For a wide class of problems $T_{\text{comp}} = \max_q (N_q T_q)$, where qth processor is responsible for N_q cells or particles, T_q is the time needed to perform one step or iteration. For the most common high performance computational systems – clusters of workstation connected by Ethernet or Myrinet – it is possible to write

$$T_{\text{comm}} = \sum_{pq} (C_{pq}/b + \delta(C_{pq})L),$$

where C_{pq} is the communication matrix, b is the effective bandgap of the net, and L is its latency.

MD with near range potentials is well suited for the parallelization, about the methods for the effective implementation of the parallel computations for the long range electrostatic interactions. Thus, the efficiency of the MD software code NAMD (Not Another Molecular Dynamics) developed by the Theoretical and Computational Biophysics group in the University of Illinois at Urbana Champaign for the biophysical problems is as high as 85% and 70% for the computer systems having 1024 and 2250 processors, respectively. Implementation of MD approach using the distributed GRID resources is more complex, first of all it is hard to provide the load balance due to the dynamic character and the heterogeneity of the computational environment. There is the computational system (MDM) that have the hardware architecture specially designed for the large scale MD computations.

Models of Water

The water molecule seems to be very simple and easy to model – two hydrogen atoms and one oxygen with overall charges of 1*e* for hydrogen atoms and -2*e* for oxygen. However, to succeed in this task one should decide where these charges are located, how far is hydrogen atoms from oxygen, whether this charges are distributed in the spherical or some other arrangement, whether it is a rigid or a flexible molecule, how it interacts with neighbors, whether the quadrupole moment of the molecule should be reproduced, etc. The many body effects are significant, but the most important are the pair interaction potentials, since as the analysis shows these interactions comprise at least about 3/4 of the total energy for the water trimer, tetramer, pentamer, water pentamer ring-like structures, and heptamer chains.

Interaction Potentials

A significant number of the models of the water molecules was suggested differing in the structure (a number of charges, their position, character of bond, account for the water molecule polarizability (the dipole moment of the water molecule is about 1.85D [1086] and the quadrupole moment is also significant) etc.), in the accuracy of the reproduction of the experimental data, and suitability for the high performance computing; the recent experiments on the X-ray Compton scattering indicate that it is necessary to include in the model the correlation of the length of the O–H bond with the structure of hydrogen bonds net. Potential parameters frequently are fitted to reproduce the water phase diagram as well as raw experimental data such as spectroscopic data (including vibration–rotation–tunneling data) obtained for small water clusters.

The literature survey shows that old water models proposed before 1980s - SPC, SPC/E, TIP3P, TIP4P - are used in the majority of the MD simulation studies.

The main model parameters are the number and value of the charges, their geometrical arrangement, and type of the interaction potentials. The first models were introduced in early 1930s by Bernal and Fowler, such as simple point charge (SPC) and transferable intermolecular potentials (TIPS). The most common ones are probably SPC and TIP3P, both having three point charges but differing in their locations (e.g., the equilibrium HOH angle is 109.47° in SPC and 104.52° in TIP3P).

In the SPC/E model, the water molecule is represented as a set of point charges with Lennard–Jones interactions between the oxygen atoms. The intermolecular potential between two water molecules is expressed as

$$U = 4\varepsilon_{OO}\left[\left(\frac{\sigma_{OO}}{r_{OO}}\right)^{12} - \left(\frac{\sigma_{OO}}{r_{OO}}\right)^{6}\right] + \frac{1}{4\pi\varepsilon_0}\sum_{i=1}^{3}\sum_{j=1}^{3}\frac{q_i q_j}{r_{ij}}, \quad \text{...(13)}$$

where σ_{OO} and ε_{OO} are the parameters of the Lennard–Jones potential, r_{ij} is the distance between the charge site i and j on two different molecules, and ε_0 is the permittivity of the free space.

Improvements of the SPC model (SPC/A and SPC/L) are considered by Glãttli et al., and "spring-on-charge" polarizable models and their modifications by Yu and van Gunsteren.

The comparison of the water models on the basis of the reproduction of the ice melting temperature and other parameters of the water phase diagram could be found in and on the basis of applicability to the dynamics of proteins and amino acids in the water solutions.

As was already noted, the classical presentation of the water molecule by a tetrahedral disposition of the two negatively and the two positively charged regions is an oversimplification and, being used as a basis for the interaction potentials, will overemphasize the angular dependence of the intermolecular potential. The more adequate potential functions represent the lone-pair region by a single negative point charge located close to or even at the molecular (oxygen) center.

TIP5P pair potential is formulated under the assumption that the water molecule is a rigid nonpolarizable tetrahedral structure presented by five point charges. Two positive charges (q_H = 0.241e) are located on hydrogen atoms at the distance 0.09572 nm from the oxygen atom forming the HOH angle equal to 104.52°. Negative charges (q_e = -qH) represent the lone pair of electrons in the plane perpendicular to the HOH plane; the angle e^-Oe^- is equal to 109.47°. The fifth interaction site introduced to prevent the overlapping of the water molecules is located on the oxygen atom and interacts via Lennard-Jones potential with the parameters σ_{OO} = 0.312 nm and ε_{OO} = 0.6694 kJ mol^{-1}.

The TIP5P is known to accurately reproduce some water anomalies such as the density anomaly at T = 277 K and atmospheric pressure and some others.

The water potentials could be very complex, for example, the so-called ASP-W potential has 72 parameters corresponding to electrostatic interactions, two-body exchange-repulsion, two-body dispersion, many-body induction.

Bulk Water

Yan et al. recently performed a detailed analysis of the water anomalies of different nature – structural, thermodynamic, dynamic. The authors used two-order parameters:

1. Translational

$$Q_t \equiv \int_0^{c_s} |g(s) - 1| ds,$$

where $s \equiv r\rho_n^{1/3}$ is scaled by the average distance between the particles, $g(s)$ is the pair correlation function, s_c is the cut-off parameter (it was set to a half of the computational domain size in the paper) and

2. Orientational

$$Q_{li} \equiv \left[\frac{4\pi}{2l+1} \sum_{m=-l}^{m=l} \left| \overline{Y}_{lm} \right|^2 \right]^{1/2} \quad ...(14)$$

A definition of the last order parameter is based on the characterization of the particle position relative to the twelve closest neighbors by the angles θ, ϕ that are used to calculate the corresponding spherical harmonics $Y_{lm}(\theta, \phi)$ and the average $\overline{Y}_{lm}$. When l = 6 the parameter Q_{6i} could serve as a measure of the orientational order in the system of particles. The necessary condition for the appearance of the anomalous structural properties is proved to be the presence of two linear scales in the interaction potential.

Detailed study of the first- and second-neigbor shells defined, respectively, by four and twelve neigbors, reported in the subsequent paper by Yan et al., showed that the anomalous decrease of the translational order upon compression occurs mainly in the second neighbor shell. The authors analyzed the oxygen–oxygen pair correlation function $g(r)$ and divided it into three regions to study the average effect of the density on the different shells. The average number of neighbors of the central molecule at a distance r was computed as

$$N(r) \equiv 4\pi n \int_0^r (r')^2 g(r') dr'.$$

The above mentioned regions of water are defined as follows:

1. First shell: $0 < r \leq r_1$
2. Second shell: $r_1 < r \leq r_2$
3. Rest water: $r > r_2$

where r_1 and r_2 depending on the temperature and pressure are implicitly given by relations $N(r_1) = 4$ and $N(r_2) = 12$.

To characterize the tetrahedrality of the first shell, the following orientational order parameter was used:

$$q_i \equiv 1 - \frac{3}{8} \sum_{j=1}^{3} \sum_{k=j+1}^{4} \left[\cos\theta_{jik} + \frac{1}{3} \right]^2, \qquad \text{...(15)}$$

as well as its average value

$$q \equiv \frac{1}{N} \sum_{i=1}^{N} q_i.$$

Here θ_{jik} is the angle formed by the central molecule i with the neighbors j and k. For the perfect tetrahedral order $q = 1$ and for the uncorrelated system $q = 0$.

The previously introduced Q_{li} (14) was averaged for $l = 6$

$$Q_6 \equiv \frac{1}{N} \sum_{i=1}^{N} Q_{6i}$$

to quantify the orientational order of the system based on the molecules in the second shell (for the uncorrelated system $Q_6 = 1/\sqrt{12}$).

The authors performed MD simulations using 512 water molecules interacting via the five-site transferable interaction potential TIP5P and found that anomalies of water are related to the alteration of the arrangement of the water molecules in the second shell.

Nonequilibrium MD simulations were used by Evans et al. to study the thermal conductivity of pure water and its dependence on various degree of the orientational and translational order. The degree of the order in water was altered by the applied electrical field. The authors used the planar heat source and heat sink within a rectangular periodic simulation box. A simple point charge flexible water molecule was used. The steady state temperature profile was established after 100,000 MD steps, after which the averaging was performed over about 300,000 MD steps. The strength of the uniform electric field was varied from 0 to 7Vnm^{-1}.

It was found that while the orientational ordering of the water dipole moments has a minor effect on the thermal conductivity, the translational ordering results in approximately threefold increase of the water thermal conductivity, making it comparable with that of the crystalline ice.

Water Confined Between Surfaces

Kumar et al. have considered the structure and behavior of pure water confined between two hydrophobic plates for several temperatures from 220 to 300K and several densities from 0.8 to 1.32 g cm^{-3}, in total over 50 state points. In MD simulations N = 512 water molecules were used that interact via TIP5P potential in the computational domain $L_x \times L_y \times L_z$, where separation between the plates $L_z = 1.1$ nm corresponds to two or three layers of the water molecules.

The interactions between the water molecules and the smooth surfaces of the plates were modeled to mimic the water interaction with solid paraffin and described as

$$U(\Delta z) = 4\varepsilon_{Ow}\left[\left(\frac{\sigma_{Ow}}{\Delta z}\right)^9 - \left(\frac{\sigma_{Ow}}{\Delta z}\right)^3\right],$$

where Δz is the distance from the oxygen atom of the water molecule to the wall and ε_{Ow} = 1.25 kJ mol^{-1}, σ_{Ow} = 0.25 nm are the parameters of the interaction potential.

The pressure along the transverse direction $P_\perp$ is determined via the total force F_w projection on the normal to the wall

$$P_\perp = \frac{F_w}{L_x L_y} = \frac{\left|\sum_{i=1}^{N} F_{i,w}\right|}{L_x L_y}.$$

Here $F_{i,w}$ is the force of the oxygen atom of the ith molecule acting on the wall; it is assumed that hydrogen atoms do not interact with the wall.

Since the repulsive interactions with the wall prevents molecules from coming close to the wall, the authors introduced the effective separation between the plates L_z', which is estimated as

$$L_z' = L_z - \frac{\sigma_{Ow} + \sigma_{OO}}{2} \approx 0.819 \text{ nm}$$

and is used to calculate the effective density $\rho \equiv Nm/L_x L_y L_z'$.

For the comparison with bulk water, Kumar et al. computed the lateral oxygen–oxygen radial distribution function (RDF):

$$g_{\parallel} = \frac{1}{\rho^2 V}\sum_{i \neq j} \delta(r - r_{ij})\left[H\left(\left|z_i - z_j\right| + \frac{\delta z}{2}\right) - H\left(\left|z_i - z_j\right| - \frac{\delta z}{2}\right)\right], \quad (16)$$

where V is the volume, r_{ij} is the distance between the ith and jth molecules parallel to the walls, z_i is the z coordinate of the oxygen atom of the ith molecule, and $\delta(x)$ is the Dirac δ function. The Heaviside functions $H(x)$ are introduced to provide that the sum is calculated for the pairs of molecules located in the same layer of thickness $\delta z = 0.1$ nm.

Computations showed that RDF at low temperature and low density exhibits distinctive maxima and minima and indicates that liquid is highly structured, with the structure parallel to the walls similar to that of bulk water. As the density increases, the liquid becomes less structured as is evident from the decreasing of the second and third peaks of $g_{\parallel}$. The lateral oxygen–oxygen RDF $g_{\parallel}(r)$ (16) is used to calculate the lateral static structure factor $S_{\parallel}(q)$, which is defined as the Fourier transform of the former:

$$S_{\parallel}(q) = \frac{1}{N}\sum_{i,j}\left\langle e^{i\mathbf{q}\cdot(\mathbf{r}_i - \mathbf{r}_j)}\right\rangle,$$

where the q vector is the inverse space vector in the xy plane and r is the projection of the position vector onto the xy plane.

The study of the diffusion of the water molecules in the different layers at high density showed that molecules in the layer adjacent to the wall move faster than molecules in the middle layer.

Kumar et al. have found the effect of confinement to be similar to the effect of the temperature elevation, with the same temperature shift about 40 K for both thermodynamic and dynamic properties. Such behavior is qualitatively explained by the alterations in the hydrogen bonds network. Since hydrogen bonds are not formed with the walls, the average number of hydrogen bonds per molecule in the confined system is smaller than that in bulk water, which is similar to having the water at higher temperature.

Wang et al. have found the confined water behavior to resemble not that of bulk water under the elevated or reduced temperature, but under higher pressure for the somewhat different case of two hydrophilic surfaces modeling (001) surface of the mineral brucite $Mg(OH)_2$. The authors performed MD simulation using SPC interaction potential. The essential distinction of this work is the consideration of the surface which was not only assumed rigid, but also was simulated using MD approach.

The relaxation of atom positions were performed in two steps. First the position of all atoms in the brucite substrate was fixed and only water subsystem was allowed to relax. Then all atoms - both in liquid water and solid brucite - were relaxed. To characterize the local structure of the hydrogen bonds network near the surface, the total number of HBs per water molecule, the number of HBs donated to and accepted from other water molecules, and the number of HBs donated to and accepted from surface OH groups were calculated. The orientational order parameter (15) was also computed with an evident modification - inclusion of the angles θ_{jik} that are formed by HBs with the surface OH groups.

It should be noted that there is two approaches to the identification of the hydrogen bonds in the MD simulations:

1. Energetic bonding criterium: two water molecules are considered bonded if their oxygen-oxygen distance is less than 3.5Å and their interaction energy is less than a specified threshold energy E_{HB} over the time interval exceeding the specified value;
2. Geometric bonding criterium: two water molecules are considered bonded if their oxygen-oxygen distance is less than 3.5Å and the O-H-O angle between molecules is less than a given threshold value θ_{HB}.

The authors found that both HB donation to the surface oxygen atoms and HB acceptance from the surface OH groups determine the structure of the first water layer that shows two-dimensional hexagonal arrangement in the plane parallel to the surface and reflects the brucite structure. The oxygen and hydrogen atomic densities and dipole orientations of the water molecules deviate from the corresponding parameters of bulk water up to the distance of about 15Å from the surface (≈5 molecular water layers). The evolution of the structure of the water layers could be described as a gradual decrease of the substrate-induced ordering. The translational ordering decays faster than the orientational one.

Systematic MD simulation of the effects of the pressure (–0.15 GPa $\leq P \leq$ 0.2 GPa), the plates separation (0.4 nm $\leq d \leq$ 1.6 nm), and the nature of the surface (hydrophobic or hydrophilic) on the structure and phase state of the confined water was performed by Giovambattista et al. Computations were conducted for 3,375 water molecules, and the temperature was fixed at 300 K using the Berendsen thermostat. Interaction of water molecules was modeled using the extended simple-point-charge pair potential (SPC/E). Structure of both hydrophobic and hydrophilic surfaces corresponded to four layers of SiO_2. The unit cell of SiO_2 was represented by a perfect tetrahedron. In the case of the hydrophobic surface, silicon and oxygen atoms interact with water molecules via a Lennard–Jones potential. The hydrophilic surface corresponding to the fully hydroxylated silica was obtained by attaching a hydrogen atom to each surface oxygen atom. The silicon and oxygen atom positions were fixed (as in the case of the hydrophobic surface), but hydrogen atoms were allowed to move in the plane without violation of the bond length. The local tetrahedral order in the water was characterized by the average coordination number and the orientational order parameter (15). Computations showed that in the case of the hydrophobic plates water molecules are pushed towards the plates and the liquid order increases with the pressure, while no effect was observed for the water orientation near the surface. No changes in the structure of the liquid with increase in the pressure were registered in the case of the hydrophilic surfaces.

As was already mention above, the effect of the wall proximity on the crystallization temperature is known from observations on the soil freezing and the water behavior in the pores of minerals. Zangi simulate using MD method the water crystallization in the narrow (several diameters of the water molecule) gap for the case of the inert surface that does not form the hydrogen bond with the water molecule. The variation of the gap width at room temperature results in the drastic (three–four orders of magnitude) variation of the diffusion coefficient that has been considered as a transition to the crystalline state named the "confinement induced freezing." Two phases of liquid water were observed. The author stressed a synergetic action of the space confinement and the electric field.

Aqueous Solutions

In the continuum description of the ions in the solution, the free energy of the solvation of an ion of radius R_i and charge q is given by the Born equation:

$$\Delta F = -\frac{q^2}{8\pi\varepsilon_0 R_i}\left(1-\frac{1}{\varepsilon}\right),$$

where ε_0 is the permittivity of the free space and ε is the dielectric constant of the medium, while the entropy of the solvation is defined as

$$\Delta S = \frac{q^2}{8\pi\varepsilon\varepsilon_0 R_i}.$$

This theory thus predicts a parabolic dependence of the free energy, entropy, and energy of solvation $E = \Delta F + T\Delta S$ on the charge for both anions and cations that is in contradiction to the experimental data. The continuum theory based on the hydrodynamic friction given by Stokes law predicts decrease of the ion mobility, irreversible to the sign of the charge, with the ion radius. In experiments, however, the reduction of mobility for both small and large ions is observed, with maximum mobility occuring at different radii for the positively and negatively charged ions. One of the proposed explanation of the low mobility of small ions – the so-called *solventberg* model – assumes that such ions (Li^+, for example) diffuse as a single entity with some solvated water molecules and thus its effective radius should be increased by the size of the water molecule (if it is assumed that solventberg includes the water molecules of the first hydration shell). This model fails, however, to explain the mobilities of other small ions such as Na^+, Cs^+, or Br^- – probably, due to the different "times of life" of the hydration shell.

Another suggested model exploit the so-called dielectric friction accounting for the inertia of the polarization of the water molecules caused by the ion. However, this model is also symmetrical with respect to the sign of charge and thus is unable to cope with differences in behavior of anions and cations. About other attempts to modify the continuum theory such as correction of the effective hydrodynamics radius and molecular theories based on the Langevin equation, see the paper by Rasaiah and Lynden-Bell. These authors performed MD simulations of ions in the aqueous solution at infinite dilution that reproduce main experimental trends. SPC/E model of water was used. The potential of the interaction between the ion and the water molecule was written similar to the water model (13) (there is a misprint in the paper – a single, not a double, sum should be present):

$$U_{iw} = 4\varepsilon_{iO}\left[\left(\frac{\sigma_{iO}}{r_{iO}}\right)^{12} - \left(\frac{\sigma_{iO}}{r_{iO}}\right)^{6}\right] + \frac{1}{4\pi\varepsilon_0}\sum_{i=1}^{3}\frac{q_i q_j}{r_{ij}}.$$

The computations were performed for a single ion and 215 water molecules at 298 K and 512 molecules at 683K. The diffusion coefficients were determined by the asymptotic slope of the mean square displacement

$$D_i = \frac{1}{6}\lim_{t\to\infty}\frac{d\left|r_i(t)-r_i(0)\right|^2}{dt}$$

and from the integral of the velocity autocorrelation function

$$D_i = \frac{1}{3}\int_0^\infty \langle v_i(t)\cdot v_i(0)\rangle dt.$$

Rasaiah and Lynden-Bell have found the following:

1. With the possible exception of the smallest solute Li, the Stokes law is valid for the movement of the neutral solutes even though their size is comparable with that of the solvent.
2. The charge asymmetry originates from the differences in the orientation of the water molecules in the first solvation shell: the oxygen atom of the water molecule in the hydration shell of the cation is located closer to the ion than the hydrogen atoms, while the reverse configuration is observed for the anion.

The authors also determined the average hydration number for different ions as a volume integral of ion–oxygen pair distribution function g_{io}

$$N_h = \rho_w \int_0^{R_h} g_{iO} 4\pi r^2 dr$$

and residence times of the water molecules in the first hydration shell that were found to decrease with size for anions and increase for the neutral solutes. The smaller diffusion coefficient of the neutral solute compared to that of the charged one is attributed to the formation of a solvent cage around the hydrophobic solute.

ICE CRYSTALLIZATION

Pure Water

The simulation of the homogeneous nucleation of the pure water is rather difficult: the first MD computations of freezing water from scratch were reported as late as 2002. These extensive computations required the integration for a period about 300 ns with the time step 1 fs. The problem is the complexity of the potential energy surface (PES) in the 6*N*-dimensional space (*N* is the number of particles in

MD computations) that has a large number of local minima, that is, numerous possible configurations of the hydrogen bonds net, and the need to simulate the rare event of the spontaneous formation of an ice nucleus.

One of the problems in crystallization simulation is the choice of the criterium to distinguish water molecules that belong to the liquid and solid phases. Since the bulk water has an average number of hydrogen bonds per molecule close to four, the instantaneous number of hydrogen bonds is not a good parameter to discriminate between ice and water. Carignano et al. suggested to monitor the time average of the number of bonds on each individual molecule. The period of time over which the average is taken should be short for the solid phase, but large enough for the liquid phase to allow molecules to diffuse. By numerical experiments the authors found the period of 20ps to be adequate. Hydrogen bonds in computations are defined by two conditions involving the distance cut-off and the angular cut-off related to the angle formed by the hydrogen, the donor oxygen, and the acceptor oxygen. The typical values of these cut-off parameters are $r_{HB} = 0.35$ nm and $\theta_{HB} = 30°$.

Briels and Tepper use another approach. They inquire whether the particle surroundings have octahedral symmetry and construct a "*recognition function*" (the order parameter)

$$\Psi = \frac{\sqrt{\langle X^2 \rangle - \langle X \rangle^2}}{\langle X^2 \rangle},$$

where

$$X(r) = \frac{x^4}{r^4} + \frac{y^4}{r^4} + \frac{z^4}{r^4} - \frac{3}{5},$$

with the averages taken over all nearest neighbors residing within a given cutoff radius and coordinates are in respect to the centre of mass of the neighbors, and not to the central particle; for "liquid" particles $\Psi > 0.5$. Since the crystalline structure will partly persist in the liquid near the interface, the authors expect that their discriminator will slightly overestimate the number of "solid" particles.

Topological properties of the PES are related to the thermodynamic and dynamic properties of the supercooled water. The authors of the cited paper studied this problem using SPE/E interaction potential in MD simulations of a system of $N = 216$ water molecules for six different densities and six temperatures in the range from 210 to 300

K. The PES can be partitioned into a number of local *basins*, which are uniquely defined by the set of points connected to a local minimum via a steepest-descent paths; the configurations corresponding to the local minima are called *inherent structures* (IS) and their energies are denoted by $E_{IS.}$ The free energy of the system F at the temperature T can be expressed as

$$F(T) = -TS_{\text{conf}}(e_{IS}(T)) + f_{\text{basin}}(e_{IS}(T),T),$$

where e_{IS} is the average value of the IS energy, S_{conf} $(e_{\text{IS}}(T))$ = $\log(\Omega(e_{\text{IS}}$ $(T)))$ is the entropic contribution arising from the number $\Omega(e_{\text{IS}})$ of basins of depth e_{IS}. The separation of the free energy into two parts reflects the separation of time scales in supercooled liquids: two-scale relaxation takes place with intrabasins dynamics being much faster than interbasins dynamics. Slowing-down of the dynamics approaching the glass transition temperature can be attributed both to the reduction of the connectivity between local minima and to the decrease in the number of available local minima.

The results of MD computations are highly sensitive to the choice of the interaction potential and could underestimate as well as overestimate the degree of the water ordering. A set of the order parameters serves as a measure of the latter. The order parameters based on the expansion of the pair correlation function in the spherical coordinates describe the geometrical properties of the bonds of neighbor molecules; an additional three-particle order parameter is related to the tetrahedral structure. The nucleation could be represented by the minimal free energy trajectory in the space of the order parameters. Two- and three-particle order parameters were used by Radhakrishnan and Trout for calculation of the nucleation energy barrier in the different environments: homogeneous liquid; water in the hydrophobic pore; water in the constant electric field; mixture of water and CO_2; water with antifreeze protein.

Simulations of the melting, of the crystallization in the spatially restricted systems, or of the crystallization of the water that is in contact with ice are less computer-intensive. Carignano et al. considered growth of ice from the supercooled water with the explicit presence of ice–vapour and water–vapour interfaces using the six-cite water model of Nada and van der Eerden. These sites are the oxygen and hydrogen atoms, the point charge located at the bisector of the angle $\angle HOH$, and two point charges on the plane perpendicular to HOH introduced to emulate the water lone pairs. The authors' computations show that the growth rate at the prismatic plane of the ice is approximately twice as large as the growth rate at the basal plane. The rough ice–water interface was registered at the prismatic plane, while smooth interface due to the layer-by-layer growth mechanism was observed at the basal plane. Contrary to the experimental data, the computations do not reveal the growth rate dependence on the supercooling.

A free surface of a solid has a quasi-liquid layer. Carignano et al. proposed a theoretical definition of the thickness of this layer to be equal to four times the standard deviation calculated from the liquid density profile $\phi_L(z)$:

$$\delta_l = 4\left[\frac{\int \phi_L(z)(z-\overline{z})^2 dz}{\int \phi_L(z)dz}\right]^{1/2},$$

where

$$\overline{z} = \frac{\int \phi_L(z)zdz}{\int \phi_L(z)dz}.$$

The thickness of this layer could be measured experimentally. The authors of the cited paper also studied the dynamics of the quasi-liquid layer. The decrease of the initially wide layer on the ice–vacuum surface is found to be linear with the time and rather fast, reaching the equilibrium value – the same as that of the layer that melted from the solid side – in about 200ps.

Swishchev and Kusalik performed MD simulations of the crystallization of the liquid water subjected to a homogeneous static electric field. The authors used TIP4P model of water; the field strengths considered were 0.1 and 0.5 V Å^{-1}, which are smaller than the average internal electric fields within the condensed phases of water. The crystal of cubic ice I_c was grown in the absence of electric field melts at 250 K. The growth of the cubic ice in the electric field instead of the hexagonal I_h is explained by the authors by the better parallel arrangement of the molecular dipoles in the diamond-like packing of I_c in comparison with the isomorphous hexagonal packing of I_h. The authors conclude that local electric fields occurring at the surfaces of various materials could play a crucial role in promoting the crystallization of the liquid water.

Muguruma et al. applied the multicanonical MC method to the bulk water system containing 64 molecules as a basic unit cell. The determination of the weight factors required about 10^5 MC sweeps and the production run about 5×10^6 sweeps. The authors obtained the water transition into the amorphous ice at 190 K and noted that for simulation of the formation of crystalline ice more accurate weight factors in the low energy region should be determined.

Aqueous Solutions

Ice nucleation in liquid water and aqueous solution is a common natural process occurring annually in mid and high latitudes. In spite of the development of the sophisticated experimental techniques enabling high-speed visual and infrared imaging of the nucleation and growth of ice in small droplets such as free-fall tubes, emulsified droplets, jet expansions, droplets hanging on wire, or acoustically suspended droplets free of wall or substrate contacts, the most detailed information on the process, especially its early stages, is provided by numerical simulation.

Vrbka and Jungwirth simulate the crystallization of the aqueous solution of NaCl that is in contact with the cubic ice I_c that occupies one-fourth of the computational domain comprising 720 water molecules; the nonpolarizable potential SPC/E was used. The system geometry was a slab with a bulk region between two air/water interfaces. The size of the system was sufficient for the study of the spontaneous formation of ice nucleus.

The time needed for the nucleus formation in the salty water turned out to be over an order of magnitude larger than that in the case of pure water. This result indicate that the effect of the presence of salt is not only the thermodynamic one (the freezing point depression), but also kinetic.

The system freezes as pure ice almost completely, with the ions being expelled into a small volume of unfrozen brine, but sometimes an ion (chloride anion) gets trapped inside the ice crystal. The instability of the crystallization front results in the formation of voids containing concentrated solution - brine. The front propagation is related to the local fluctuations of the salt concentration: local reduction of the salt density close to the crystallization front is followed by building a new ice layer. Note that the problem considered is of a considerable interest for geophysics: an increase of the solution density under the surface ice significantly influences the ocean circulation.

Transport in Small Pores

MD simulations are used to study water and small molecules in the pores of different nature. The most popular objects for investigation are carbon nanotubes (CNs) and channel formed in the cellular membranes by the integral proteins such as aquaporin and gramicidin. CNs are of interest by themselves for various technological applications and also as a potential model of the biological channels. The discussion of the simulation of the metabolically pumped ion channels and their gating is beyond the scope of the present work.

As was already mentioned, integral proteins from the aquaporin family form in the cellular membrane tetramers with a functionally independent channel in each monomer that is permeable for water but not for the charged molecules and ions, while the fifth channel in the center of the tetramer is permeable for both water and charged particles.

Experimentally osmotically driven fluxes through biological channels are measured by mounting these channels into the lyposomes and monitoring the volume change of the latter in response to the concentration of the environment media. A major experimental challenge is the need to determine exactly the protein-to-lipid composition, that is, the density of the water channels to be able to extract the single-channel properties from the measurements of the membrane permeability. Evidently, these difficulties are not encountered in the MD simulation study.

The properties of the small pore are characterized by two parameters: osmotic permeability p_f and diffusion permeability p_d. The determination of the latter from MD simulations is trivial. It could be found by counting the water molecules passing from one side of the channel to the other as

$$p_d = \frac{V_w}{N_A} q_0 = \upsilon_w,$$

where V_w is the molar volume of the water, N_A is Avogadro number, q_0 is the unidirectional permeation rate, υ_w is the volume of a single water molecule.

In the dilute solutions, the total water flux j_w through a single channel is linearly proportional to the solute concentration difference Δc_s: $j_w = p_f \Delta c_s$. The second permeability parameter, p_d, corresponds to the case of $\Delta c_s = 0$.

To determine p_f from MD computations, one should produce difference of osmotic or hydrostatic pressures on the two sides of the membrane containing the pore. Zhu et al. proposed to use periodic system in which the unit cell is replicated in all three dimensions and thus water layers and membranes alternate in the direction z normal to the membrane plane. The authors define three zones (I, II, and III) for each water layer, with zone III being the center of the layer. A constant force f acting on all molecules in the zone III along z-direction is added to generate a pressure gradient in this region, thus providing the pressure difference between the zones I and II: $\Delta P = P_{I} - P_{II} = Nf/A$, where N is the number of water molecules in the zone III and A is the area of the membrane. Therefore, the net water flux across the membrane is induced and p_f could be calculated from j_w and ΔP.

The computations performed by Zhu et al. using aquaporin molecule in the lipid bilayer composed from palmitoyl-oleyoyl-phosphatidyl-ethanolamine (POPE) molecules (the complete system, including water, contains 81,065 atoms) gave the p_f/p_d ratio about 12.

This approach was used later by Jensen and Mouritsen to study the single-channel water permeability of *Escherichia coli* aquaporin AqpZ and aquaglyceroporin GlpF, and by Mamonov et al. to compute the water and deuterium oxide permeabilities through the channel formed by aquaporin. The authors used the protein structure from the Protein Data Bank and built the tetramer that was pre-equilibrated in 177 POPE molecules and solvated in 15,079 water molecules; TIP3P interaction potential was used for the water. Four different values of force were used in the method of Zhu et al. to produce four pressure differences. Computations were performed using earlier mentioned NAMD package.

Zhu et al. suggested for the determination of the permeabilities in channels where the so-called *water file* – the chain of water molecules with the highly correlated movement along the channel – exists to apply the continuous-time random-walk model developed by Berezhkovskii and Hummer for the water transport through the carbon nanotube. The model is developed under the assumption that the channel is always filled by N water molecules that move together in hops, with the left and right hopping rates k_l and k_r being equal to k_0 in equilibrium.

The complete permeation of the single water molecule from one side of the channel to the other requires at least $N + 1$ hops, thus it is expected that the number of unidirectional permeation events per unit time will be given by $q_0 = k_0/(N+1)$, then the diffusion permeability $p_d = \upsilon_w k_0/(N+1)$.

In the measurements of p_f, the net water flux is induced by the different solute concentrations on the two sides of the membrane and thus instead of $k_l = k_r$ one should have

$$\frac{k_r}{k_l}\exp\left(\frac{-\Delta\mu}{kT}\right).$$

Since under physiological conditions the difference in the chemical potential $\Delta\mu$ is small compared with kT, it is reasonable to use a linear approximation and write

$$k_r = k_0\left(1+\alpha\frac{-\Delta\mu}{kT}\right), \quad k_l = k_0\left(1+\beta\frac{-\Delta\mu}{kT}\right),$$

where $\alpha = -\beta$ for the symmetrical channel. Then the net water flux is

$$j_w = \frac{1}{N_A}(k_r - k_l) = \frac{k_0(\alpha-\beta)}{N_A}\frac{-\Delta\mu}{kT}.$$

Hashido et al. argue that a scalar permeability could not describe correctly water transport through the channels formed by aquaporin and similar integral membrane proteins and osmotic permeability matrix should be introduced to account for contributions of all local regions of the channel on the water transport. Diagonal elements of this matrix named p_f-matrix are equivalent to the local permeability at each region of the channel, and off-diagonal elements represent correlated motions of water molecules in different regions. Averaging both diagonal and off-diagonal elements recovers the scalar permeability for the entire channel. The authors have performed MD computations to construct the p_f-matrix for five members from the aquaporin family.

The osmotic transport in the cylindrical pore with a neutral wall (no interactions with molecules) was investigated by Kim et al. and the water transfer through the lipid bilayer by Berendsen and Marrink.

It is stressed in the latter paper that a characteristic time of the water molecule transport is rather high (about 100ps), thus MD should be used not for the direct simulation of the transport, but rather for calculation (using MD distribution function of water molecules) of the coefficients needed to close the macroscopic model of the water transfer. The authors considered the water transport through the membrane consisting of dipalmitoyl phosphatidycholine (DPPC). Membrane was assumed to be in the liquid crystalline state. The computations were performed for a rectangular periodic box at constant pressure and constant temperature. Never a cluster of two or more molecules inside the membrane was observed. Since the quantitative study of the water permeation process in an equilibrium simulation is impossible, the authors considered it as a diffusion in an external field described by free energy profile or potential of mean force. The average velocity u_i of the ith species (here water) and hence its flux J_i are proportional to the gradient of the thermodynamic potential:

$$u_i = -\frac{1}{\zeta_i}\nabla\mu_i, \quad J_i = -\frac{c_i}{\zeta_i}\nabla\mu_i,$$

where the friction coefficient ζ_i is related to the diffusion coefficient $\zeta_i = R_g T/D_i$. In an ideal solution $\mu = \mu_0 + R_g T \ln c$, so the above equation is just the Fick law $J = -D\nabla c$. In the steady state the flux is constant in every point of the membrane, thus the states on two membrane sides $z = z_1$ and $z = z_2$ are related as

$$\Delta\mu_i = \mu_i(z_2) - \mu_i(z_1) = -J_i R_g T \int_{z_1}^{z_2} \frac{dz}{c_i(z)D_i(z)}.$$

The permeation resistance (that is directly related to the permeability coefficient measured experimentally) is defined as

$$R_i^p = c_i \int_{z_1}^{z_2} \frac{dz}{c_i^{\text{equil}}(z)D_i(z)},$$

where the equilibrium water distribution $c_i^{\text{equil}}(z)$ is determined from MD computations.

Kim et al. performed comparison of the permeability and the diffusive drag coefficient obtained from MD computations with the value that follow from the continuum description provided by the classical fluid dynamics, introducing two coefficients

$$G_w = \frac{p_f}{p_{f(\text{Poiseuille})}} \quad \text{and} \quad F_w = \frac{p_d}{p_{d(\text{Fick})}},$$

where the classical coefficients are expressed as

$$p_{f(\text{Poiseuille})} = \frac{\pi r_{\text{pore}}^4}{8\eta L_{\text{pore}}}$$

and

$$p_{d(\text{Fick})} = \frac{\pi r_p^2 D}{L_{\text{pore}}}$$

where the viscosity η is obtained as the free viscosity η_0 corrected by the so called *wall factor* in the following way:

$$\eta = \eta \frac{r_{\text{pore}}}{r_{\text{pore}} + k\overline{x}}.$$

Here $\overline{x}$ is the mean free path of the particle and k is the slip factor varying from 1 (no slip) up to 8.

The results of computations showed that G_w could be as low as about 0.3 and F_w is smaller than 0.1.

The MD study by Kalra et al. of the water transport through the carbon nanotube using hard-sphere particles to model water molecules confirmed the existence of the continuous single file of water (a hydrogen-bonded *water wire*) and explained the paradoxial from the point of the classical fluid mechanics weak dependence of the overall conductance on the channel length. The single water file behaves as rod that slips through the pore faster than is according to the unrestricted diffusion. The absence of the length dependence follows from the dominance of the entry and exit effects over the movement through the pore itself. However, the carbon nanotube hardly could be considered as a good model of the biological channels due to the much more complex structure of the latter that could contain cavities and extremely nonuniform distribution of the charged protein residues along the channel.

The mechanisms of the proton and hydroxide migration along the linear water chain in the channel of aquaglyceroporin GlpF from *E. coli* was studied by Jensen et al. using hybrid quantum–classical simulations (QM/MD) that combines the first-principle ab initio MD method of Car and Parinello (CPMD) based on the density functional theory and the classical MD method based on the force-field potentials. The system was partitioned into QM and MM fragments that were treated by the corresponding methods. For the QM regions, interactions with distant MM atoms were processed using a multi- pole expansion. The authors found that the single water file containing eigth molecules is unable to support the Grotthuss-type mechanism of the proton transport.

Protein Denaturation

The use of the lattice models of macromolecules reduces requirements to the computer resources due to the radical reduction of the number of the possible conformations of the macromolecule. However, these model introduce a priori the space anisotropy; thus lately the most attention is paid to the *off-lattice* models. Both water molecules and parts of the macromolecule are represented by the spheres of different diameters; the interaction between the fragments *i* and *j* of the macromolecule that are not neighbors is described by the short range potential. Frequently, the cutoff Lennard–Jones potential

$$V_{ij}(r) = \begin{cases} \varepsilon\left[(\sigma_{ij}/r)^{12} - 2(\sigma_{ij}/r)^6 + \upsilon_{ij}\right], & r \le R_{ij} \\ 0, & r > R_{ij} \end{cases}$$

is used. Here a shift υ_{ij} provides the continuity of the potential at $r = R_{ij}$, while the cutoff radius depends on the properties of the elements i and j.

To monitor the particles' position, the algorithm of "random lattices" could be used.

The main approaches to the construction of the hybrid models are as follows:

1. A representation of the space fragment of the macromolecule (e.g., the phosphate or methyl group) as a single particle or a pair of particles with some effective properties.
2. Combining the discrete (atomistic) description of the macromolecule with continuum description of the solvent (water in our case). The review of such "implicit" treatment of the water focused on the simulation of the structure and function of the biological membranes, including continuum and particle-based - tethered, nontethered, nontethered solvated, and nontethered implicit solvent - models has been published recently by Brannigan et al.

In the HP model of proteins (as well as in its off-lattice generalization - AB model) amino acid residues are reduced to the point entities that could be either polar (P, represented by the charge or dipole) or nonpolar (H). Hydrophobicity is described as the intention to reduce the area of the surface in contact with water and is related to the reordering of the neighbor molecules into the energetically more attractive structure similar to the ice. This effect could be described formally as an attractive interaction of the nonpolar amino acid residues. A hydrophobic molecule in solution displays water molecules that at low temperatures form an ice-like cage around the hydrophobic molecule, producing a structure that is energetically more favorable than bulk water. The system tries to minimize free energy by maximizing the number of cages - exposing as many hydrophobic molecules as possible. This is the cold denaturation of proteins. At higher temperatures water molecules could not build cages; disordered water molecules around hydrophobic aminoacid residues of the protein are energetically less favorable than bulk water; thus proteins try to hide their hydrophobic parts inside their core. De Los Rios and Caldarelli model protein as a self-avoiding walk (SAW) on a lattice. A variable with q states (labeled from 0 to $q - 1$) is attributed to every lattice site nonoccupied by the polymer. The state $q = 0$ is associated to the cage configuration, which is energetically favorable when the water molecule is in contact with the polymer and the rest states $q = 1, \ldots, q - 1$ are associated to disordered, unfavorable, configurations. The Hamiltonian of the system is written as

$$H = \sum_{i=1}^{N} \sum (-J\delta_{s_j,0} + K(1-\delta_{s_j,0})).$$

Here the first sum run over N monomers of the polymer while the second one run over the water occupied nearest neigbors of each monomer, J and K are positive constants representing the energy of the cage and of the disordered configurations; there are no monomer–monomer interactions.

The partition function of the system is written as

$$Z_N = \sum_C Z_N(C),$$

where $Z_N(C)$ is the partition function associated to a single configuration C. Note that the maximum number of water sites in contact with the polymer is $M = 2(d-1)N + 2$ for the hypercube lattice in d-dimensional space; the real number of contacts is, of course, smaller than M. A particular water site could be counted more than once. The configuration partition function is expressed as

$$Z_N = q^{n_0(C)} \prod_{l=1}^{z} Y_l(C),$$

where

$$Y_l(C) = ((q-1)e^{-\beta lK} + e^{-\beta lJ})_l^n(C)$$

and $n_l(C)$ is the number of water sites with l polymer contacts, z is the coordination number of the lattice $\beta = 1/k_B T$.

De Los Rios and Caldarelli analyzed the thermodynamic behavior of the system on the two-dimensional lattice $d = 2$ using the exact enumeration technique for polymers of length up to $N = 25$. The so-called Manhattan lattice was used (it is a 2D lattice on which rows (columns) are alternatively left/right (up/down) oriented). The authors state that simulations using Manhatten lattices give results similar to those obtained on the regular Euclidian lattices. Only bimodal case $q = 2$ was considered.

The authors computed the energy of the monomer–monomer interaction ε using partition functions for two configurations C and C' that differ in one monomer–monomer contact:

$$\varepsilon = -\frac{1}{\beta} \ln \frac{Z(C')}{Z(C')}$$

and found the low and high temperature limits of this energy

$$\varepsilon \approx 2J, \quad T \to 0$$

$$\varepsilon \approx -2\left(K - \frac{J+K}{q}\right), \quad T \to \infty.$$

Thus the energy of the effective monomer–monomer interaction is repulsive for low temperatures.

The explicit description of the water is necessary, in particular, for the correct simulation of the cold protein denaturation.

Complete denaturation (cold unfolding) of globular proteins upon decreasing temperature is similar to the abrupt swelling transition at low temperatures of the so-called thermoreversible homopolymers. Buzano et al. used dynamic Monte Carlo simulations to study swelling of a hydrophobic polymer in the aqueous solutions. The water was described by a 2D model on a triangular lattice using the Mercedes-Benz-type model for the interaction potentials. The polymer was represented by a self-avoiding walk on the same lattice. The authors were able to observe the low-temperature-induced swelling transition at high pressure only and formulate an important question: whether the lattice water models overestimate the stability of the ordered phase (ice)?

The simple model of the hydrophobic interaction does not reproduce the pressure effect on the cold protein denaturation; a correct simulation of this process at high pressure requires the account of the water fluctuations. In the cited paper the nonpolar heteropolymer is considered. Its interaction with the water (a partial ordering of the water molecules around the protein) is described by the Hamiltonian:

$$H_p = J_r n_{HB}\left(n_{\max} - \sum_{i,j} n_i n_j\right),$$

where the parameter J_r defines the strength of the repulsive interaction, $n_{HB} \equiv N_{HB}/N_w$ is the numerical density of the hydrogen bonds, which increase the system volume, N_w is the number of water molecules, and n_{max} is the maximal number of contacts of the amino acid residues with each other; $n_i = 1$ if ith node of the lattice is occupied by the residue and 0 otherwise. The expression in round braces is the measure of the protein compactness.

The Hamiltonian of the interactions between the water molecules is written as $H_{HB} = -J\Sigma_{i,j}\delta_{\sigma_i,\sigma_j}$; the variable σ_i specify the orientation of the water molecule – the formation of the hydrogen bond between the neighbor molecules being feasible if $\sigma_i = \sigma_j$. Enthalpy for the system "protein in the water bath" is $W = H_p + H_{HB} + PV$. The authors used the multicanonic algorithm of the Monte Carlo method to compute the two-parametric (the number of the hydrogen bonds and the number of the contacts of the amino acid residues) density of states. Then the dependence of the average number of the residue-residue contacts $\bar{N}_c$ on the temperature and on the pressure could be found:

$$\bar{N}_c = \sum_{N_{HB}} \sum_{N_c} N_c g(N_{HB}, N_c) \frac{\exp - \left(\frac{W(N_{HB}, N_C)}{T} \right)}{Z},$$

where Z is the partition function. The authors performed computations for the system of a protein with 17 nonpolar residues and 383 water molecules with periodic boundary conditions.

The results have been used to construct the phase diagram of the protein- water system in the p-T plane. The condition of the denaturation was formulated as $\bar{N}_c / n_{max} \geq \alpha_d$ (it was assumed $\alpha_d = 0.96$ in the paper).

The authors stress that while they were able to reproduce the experimentally determined thermodynamics of the cold denaturation the key element of which is the pressure effect on the water density, the analysis of the kinetics of this process will require more detailed models accounting for the actual number of hydrogen bonds for each water molecule and not just the average number of bonds.

The free energy of the protein denaturation and its dependence on the pH of the solution was considered by Kundrotas and Karshikov. The authors, in contrast to many other models that deal with a single space distribution of the protein charges, consider the protein denaturated state where titratable groups are distributed quasi randomly, obeying the protein sequence restrictions. The protein molecule was represented by a chain of N elements (titratable sites) that are located on the sphere that comprises the material of denaturated protein. The allowed positions of elements are predetermined by a set of uniformly distributed points on the surface forming a spherical grid. The authors have also computed the desolvation energy and electrostatic interaction between proteins.

There are other approaches to account for the water explicitly, for example, the model of the solvent-phantom, where the interaction of the water molecules with the macromolecules is considered while the interactions between the water molecules are neglected.

AFPs and Their Binding to Ice

The mechanisms of ice growth inhibition by AFPs and AFGPs are difficult objects for the experimental study. In contrast to enzymatic proteins that can often crystallize along with their ligands, there is no direct way to study the molecular interactions between antifreeze proteins and the surface of the growing ice crystals. Even more difficult is the investigation of the antifreeze protein effect on the ice nucleation. Some hopes are related to new sophisticated techniques such as "double oil layer microsized ice crystallization technique" that allows to minimize the influence of the container on the measurements. This approach provides a way to measure the value of the free energy barriers associated with different dynamic steps of nucleation, including the incorporation of the water molecules onto the surface of ice nuclei at the kink sites and the inhibition of this process as a result of adsorption of additive or impurity. The authors of the cited paper were able for the first time to explain that mechanism of the nucleation inhibition on the ice crystal and on the surface of the foreign dust particles demonstrate some common features. The authors stress that it is important to distinguish between the nucleation induction time t_{nucl} and the time t_g needed for ice crystals to grow from the critical size to the observable size in experimental investigation of the nucleation processes. The often used induction time is the sum of these two characteristic times and is close to the nucleation time only if $t_g \ll t_{nucl}$. Of course, this relation is often valid since the energy barrier of nucleation is higher than that of growth, but exceptions are possible. Du et al. also report that similarly to surfacant molecules, type III AFP molecules accumulate and self-organize on the surface of water, which is attributed to the presense of both hydrophobic and hydrophilic parts.

The free energy $\Delta G_{AFP\text{-}ice}$ of AFP–ice binding could be estimated from the experimental data on the antifreeze activity. This energy proves to be rather small, being the difference of two large quantities – the energy of interaction between AFP and ice binding surfaces in vacuum and the energy of hydration of these surfaces. The authors of the just cited paper warn that it is difficult to accurately obtain this value in computations for two reasons: (1) the uncertainty in the interaction potentials and (2) the need to include a very large number of water molecules (thousands) into the system of AFP and ice to ensure complete hydration. The reversible process of the AFP binding could be formally written as AFP + ice ⇔ AFP:ice with k_+ and k_-

being the rates of the direct and inverse reactions related through the fraction θ of the ice surface covered by AFP:

$$k_+(1-\theta)C_{AFP} = k_-\theta$$

and whose ratio gives the association constant $K_a \equiv k_+/k_-$. The surface coverage could be thus expressed as

$$\theta = \frac{K_a C_{AFP}}{1+K_a C_{AFP}} = \frac{C_{AFP}}{K_d + C_{AFP}},$$

where the dissociation constant $K_d \equiv 1/K_a$ was introduced. Defining the fractional hysteresis activity ΔT_F as the ratio of the thermal hysteresis temperature at the given concentration of AFP to the maximal thermal hysteresis temperature $\Delta T_F \equiv \Delta T_C/\Delta T_{max}$ and assuming that it is a function of coverage $\Delta T_F = \psi(\theta)$, the authors note that experimental data could be reasonably fit for the choice $\psi(x) = x$. To get the needed estimate for $\Delta G_{AFP\text{-}ice}$ it remains to assume that it is related to K_d by the Arrhenius equation $\Delta G_{AFP\text{-}ice} = RT \ln K_d$.

It should be, however, stressed that not the energy of AFP binding to ice matters, but the *difference* between its interaction energy with ice and liquid water.

Still, up to date numerical simulation is the main source of the information on the AFPs and their function.

Jorov et al. used Monte Carlo for the minimization of the energy of the complexes AFP–ice in vacuum accounting for nonbounded interactions, hydrogen bonds, and the hydration potential for proteins. They computed MC trajectories from many different positions of AFP at different ice samples. If AFP helix is considered as a rigid body, its position and orientation are characterized by six variables corresponding to three translational and three rotational degrees of freedom, all of which were varied in MC minimization. Mutual disposition of AFP in the bounding state and ice when the helix axis of AFP is approximately parallel to the ice surface could be described by three parameters - distance from this axis to the surface and two angles (between the helical axis and the *a* axis on the basal plane of ice and the angle, characterizing rotation of the AFP around the helical axis). Variation of the latter angle will cause different AFP faces to approach the ice surface. The authors found three different binding configurations with very close values of energy and concluded that geometric complimentary alone does not explain the specificity of the AFP–ice association.

Consideration of AFP–ice complex in water using explicit water models requires huge execution time; Jorov et al. refer to the computations made by Madura et al. of binding the winter flounder AFP to ice that took months at supercomputer. MC minimization turned out to be considerably less computationally demanding. The authors found that convergence of the minimization process can be greatly improved by increasing the dielectric constant, with only minor effect on the optimal orientation of AFP on ice.

MD method was used to study the state of the antifreeze protein extracted from *Pseudopleuronectes americanus* in vacuum, in the aqueous solution, and on the surface of ice. The studied protein has 37 residues with 11 residue repeats. The AMBER all-atom force field for proteins and TIP3P potential for water were used. Computations were performed at the constant temperature of 300 K. This choice rather than temperature 273 K was motivated by the computation complexity: it is easier to provide appropriate sampling of the phase space at higher temperature.

The most favorable for the inhibition action of AFP crystallographic planes of ice were established; the relative role of the hydrogen bonds and Van der Waals forces as well as mutation effects (removing of one of the peptide groups) were considered. The authors identified four distinct ice-binding regions on the AFP surface that could bond to ice simultaneously and determined the number of the hydrogen bonds each of these regions could form with the ice surface.

This approach was also used along with the energy minimization to study the system *ice–AFP* in vacuum and in the water. The aim was to find out what peptide groups are responsible for the suppression of ice growth. Contrary to the existing opinion, the significant role of the hydrophobic amino acid residues was demonstrated. The effects of the solvent in the system "water–ice–AFP" were investigated by Kai et al.

Chen and Jia used a combination of the numerical approaches (molecular docking, energy minimization, molecular dynamics) to find the best (i.e., the most energetically favorable) ice-binding surface of the fish type III AFP, exploiting no prior knowledge of known ice-binding residues. They considered two arrangements of ice: "prism" ice representing hexagonal ice I_h and "random" ice, in which the water molecules are randomly positioned, still providing the same average density. The fish AFP considered consists of 66 residues; its structure is known from X-ray crystallography and NMR studies. MD simulation were performed after the energy minimization for the AFP–ice complex; no constraints were imposed at the MD stage, except holding the ice oxygen atom position fixed. The authors determined the patch of the protein surface that is likely to bind to ice; ice-binding site of Type III AFP turned to be larger than previously supported. Participation of the peripheral hydrophobic residues support the opinion that such interactions play an important role in AFP binding to ice.

Jorov et al. presented a simple qualitative argument while it should be so: "rigid" water molecules incorporated into ice hardly can provide contacts with AFP that are more optimal in terms of van der Waals, electrostatic, and hydrogen bond energy in comparison with "soft" molecules of bulk water that leaves hydrophobic interactions as the most probably driving force of AFP–ice association. Water molecules around hydrophobic groups maximize the number of the hydrogen bonds by forming enthropically unfavorable cage-like structures. Hydrophobic groups are getting together to minimize the total surface exposed to water. Interacting with ice, AFP hydrophobic groups dip into the grooves on the surface of the ice crystal, thus releasing the caged water to the bulk phase.

Analysis of the structure–activity relationships of different AFPs from both experimental data (mainly for the fish AFPs) and numerical simulations allowed to formulate a number of necessary conditions of the antifreeze activity:

1. AFP should be water-soluble
2. The AFP ice-binding surface should have hydrophobic groups
3. AFP should have rigid backbone conformation to prevent hydrophobic collapse of these groups in water
4. The AFP ice-binding surface should be complimentary to the ice surface
5. The AFP hydrophobic face should not be self-complimentary to prevent self-aggregation.

The above criteria are in a sense static ones and do not cover the important issue of the AFP molecule behavior in respect to the crystallization front when the latter is propagating. Two points of view have been declared: (1) adsorbed AFP molecules are in dynamical equilibrium with the free ones and (2) the absorption is permanent. Knight and Wierzbicki presented strong arguments for the latter option. If a dynamic equilibrium occurs, every time when AFP molecular is (temporaly) detached, ice front could irreversably advance some distance into the supercooled water, since there is no energetic barrier to the new layer nucleation, except the case of growth on the basal plane.

To fulfill its function, permanently absorbed AFP molecule should resist both an engulfment by the propagating ice crystallization front and pushing forward from the interface. Knight and Wierzbicki suggested a quantitative measure of AFP molecule suitability for the task, assuming it can be approximated by a spherical particle and characterized by its interfacial energies with water and ice γ_{pw} and γ_{pi} and introducing notions of "icephobic" and "icephilic" properties that are determined by the sign of the difference (γ_{pw} - γ_{pi}). The authors, however, warned that this construction is based on the sharp water/ice interface that in reality could represent a transition zone of finite width. The unacceptance of the nonpermanent absorption is probably the reason why AFP molecules are rather large: other circumstances being equal, the molecule with a smaller binding site to ice will be more easily detached from the surface by thermal fluctuations.

Cryoprotectors

The comparative analysis of the disaccharides as potential cryoprotectors was done using MD; it was showed that trehalose (α-*D*-glycopyranosyl-α-*D*-glycopyranosid) provides bonding of the greater number of water molecules and more homogeneous solution, reducing thus the probability of nucleation. The ability of trehalose to destroy the tetrahedral structure of the water and prevent ice crystallization is confirmed by the study of the thermodynamic properties of disaccharide solution using DSC method.

The properties of the lipid membranes are determined by the physical and chemical processes with spatial scales from atomistic to mesoscale (microns) in the wide range of the characteristic times. Different models focus on different spatial and temporal scales; ideally, these models should be coupled (multi-scale modeling), which require the specification of the two-way information flow between the levels of description. The review of the MD simulation for the bilipid membranes performed before 2000 could be found; later models ware considered. The model lipid systems (a monolayer at the air–water interface called Langmuir layer, planar bilayers, assemblies of bilayers in the lamellar phase), in addition to their main purposestudy of the properties of the biological membranescould serve, due to their relative simplicity, as ideal objects for the development and verification of the simulation methods. Recently, a comparison was made of the oligomer chains in the isolated state and in the membrane structure; computations of the order parameter for the bonds C–C and C–H relative to the principal axes of the tensor of inertia of the macromolecule showed that the intrinsic chain properties defined by its structure are conserved in the membrane. Simulation of the membrane protein behavior is performed. At the present time the hottest topic is the simulation of the ion channels; that is, the possibility of using of the carbon nanotubes for the emulation of the biological transport processes and collective dynamics of ion channels are studied.

MD is used to study the interaction of the cryoprotectors (trehalose, glycose, methanol, DMSO) with model membrane that differ in lipid saturation in the isothermal conditions; it was proved that small organic molecules reduce the membrane thickness, increase the area per lipid molecule, and increase the membrane permeability. The protective mechanism of saccharides including sucrose and trehalose acts through the substitution of the water molecules in the hydrogen bonds with the lipid heads and stabilization of the lipid bilayer structure - a considerable shift of the temperature of the main phase transition. Qualitatively the effect is described by the order parameter:

$$S_{CD} = -\left(\frac{2}{3}S_{xx} + \frac{1}{3}S_{yy}\right),$$

where $S_{\alpha\beta} = (3 \cos \Theta_\alpha \cos \Theta_\beta - \delta_{\alpha\beta})$ $(\alpha, \beta = x, y, z)$, $\cos \Theta_\alpha = e_\alpha e_\beta$, e_z - a unit vector in the direction z in the laboratory coordinate system, and e_α is the unit vector in the local system defined by the three successive carbon atoms C_{i-1}, C_i, C_{i+1} in the tail of the lipid molecule, $e = r_{i+1} - r_{i+1}$. This parameter that could be found experimentally using NMR method characterizes the orientation of the lipid tails with respect to the normal to the lipid bilayer.

At low temperatures the ordering of the tails increases, which assist the transition to the gel phase. Note that the problem of the interaction of the low molecular weight organic compounds with the lipid membrane arises in a quite different environment - wine production: the interaction of the alcohol molecules with the yeast membrane alters the membrane protein conformations and interferes with their function (arrests fermentation)).

The aforementioned models treat macromolecules as the flexible chains. Sometime more simple models (so called spin models) could be used that describe the orientation of the amphiphilic molecule as a whole entity. Finally, it is possible to use a continuum description of the amphiphils through the function of the free energy depending on the small number of parameters, for example, local concentrations, models of the Landau–Ginzburg type.

8

Macroscopic Models

A two-dimensional problem of an interaction of the isolated cell in the solution with the ice crystallization front was considered by Carin and Jaeger using the arbitrary Lagrangian–Eulerian method on the regular and unstructured triangular grids and by Mao et al. using the interface fitting method on the fixed Cartesian grids. A trapping of the cell by the moving crystallization front was considered in the first papers while Mao et al. simulate the behavior of the cylindrical crystallization front that is coaxial to the cell located in the center (to resolve the front, a rather large computational grid - up to 250,000 elements - was needed). The change of the cell volume due to dehydration was taken into account in the cited papers but its shape (cylindrical in the two-dimensional formulation) was deemed invariant.

Freezing of Cell Suspensions

Heat Transfer During Cryopreservation of Cells in Solution

Frequently, because of the smallness of samples used for cryopreservation by classical slow cooling or by vitrification, the temperature throughout the cell suspension could be assumed constant, thus zero-dimensional ("lumped") models of heat transfer provide a sufficient accuracy. Probably the only exception is the case of the ultra-fast cooling techniques using OHP for freezing small samples in the ultra-thin straw (UTS), where assumption of the temperature uniformity is not valid. Jiao et al. studied the heat transfer in this system starting with the bioheat equation in cylindrical coordinates:

$$\frac{1}{r}\frac{\partial}{\partial r}\left(\lambda r\frac{\partial T}{\partial r}\right)+\frac{\partial}{\partial z}\left(\lambda\frac{\partial T}{\partial z}\right)+\dot{q}=\rho c_p\frac{\partial T}{\partial t} \qquad ...(1)$$

where λ is the thermal conductivity, c_p is the specific heat, and $\dot{q}$ is the metabolic heat source. The effect of the latent heat release is assumed to be zero, since the volume ratio of the ice quantity to the maximum crystallizable ice x is typically less than 1×10^{-3}. The straw diameter (100 μm) is much smaller that its length ($\approx$5 cm), thus it is possible to disregard heat transfer along the straw axis z and, additionally assuming that thermal properties of the sample inside UTS to be uniform and temperature-independent, to get a simplified form of (1) as

$$\frac{1}{r}\frac{\partial}{\partial r}\left(r\frac{\partial T}{\partial r}\right)=\frac{1}{\alpha}\frac{\partial T}{\partial t}, \qquad ...(2)$$

where the thermal diffusivity α is introduced.

For the initial temperature T_0 and the final temperature T_∞, the transient temperature distribution of the dimensionless temperature

$$\theta^*=\frac{T-T_\infty}{T_0-T_\infty}$$

is written as

$$\theta^*=\sum_{n=1}^{\infty}C_n\exp(-\zeta_n^2 Fo)J_0\left(\zeta_n\frac{r}{r_0}\right), \qquad ...(3)$$

where

$$Fo=\frac{\alpha t}{r_0^2} \quad\text{and}\quad C_n=\frac{2}{\zeta_n}\frac{J_1(\zeta_n)}{J_o^2(\zeta_n)+J_1^2(\zeta_n)}.$$

Here J_0 and J_1 are Bessel functions and ζ_n are the positive roots of the transcendental equation:

$$\zeta_n\frac{J_1(\zeta_n)}{J_0(\zeta_n)}=\frac{2hr_0}{k},$$

where h is the heat transfer coefficient – the parameter that in some sense characterizes the quality of the cryopreservation system.

The evolution of the volume fraction of ice x is governed by the equation

$$\frac{dx}{dt}=ka_1(x)(T_m-T)\exp(-Q/RT), \qquad ...(4)$$

where

$$k=\frac{L}{\pi\lambda^2 vT_m r_f}, \quad a_1(x)=x^{2/3}(1-x),$$

T_m is the temperature at the end of the freezing process, Q is the activation energy, R is the gas constant, L is the latent heat, r_f is the radius of ice when x is equal to 1, λ is the thickness of the transition layer between liquid and solid ("mushy zone"), and ν is the kinematic viscosity.

By integration of both sides of (4), Jiao et al. obtained the relation that implicitly defines x

$$A(x) = k\int_0^t (T_m - T)\exp(-Q/RT)dt,$$

where $T = T_\infty + \theta^*(T_0 - T_\infty)$ and

$$A(x) = \int_0^x \frac{dx}{x^{2/3}(1-x)} = -\ln(1-x^{1/3}) + \frac{1}{2}\ln(1+x^{1/3}+x^{2/3})$$

$$+\sqrt{3}\arctan\left[\sqrt{3}x^{1/3}/(2+x^{1/3})\right].$$

Coupled Heat and Mass Transfer During Cryopreservation of Cells in Solution

Liu et al. applied the level set approach developed by Osher and Sethian to the simulation of freezing of biological cell suspensions. Because of instability of the crystallization front propagation into the supercooled solution, dendritic growth often occurs that is accompanied by the so-called microsegregation and macrosegregation – the continuous redistribution of solute. The complete description of the growth is extremely difficult due to the wide range of length and time scales that span many orders of magnitude – from the scales of molecular processes ($\approx 10^{-9}$ m and 10^{-12}s) to macroscopic scales of the equipment.

A transient zone between the liquid and solid in dendritic growth where both liquid and solid phases are present has a finite width and is usually referred to as *mushy zone*. The methods for moving boundary problems are divided into two groups:

1. Front tracking methods
2. Front capturing methods

The former are Lagrangian in nature while the latter are based on the Eulerian description. The Eulerian approaches are advantages in applications involving complex geometry and/or three-dimensional problems. The level set method along with a number of other powerful approaches belong to the second group.

If the details of the crystallization front geometry and the structure of the mushy zone are not of great interest, the so-called *one phase* description based on volume averaging or mixture theory provides a sufficient accuracy. In this approach there is no need to explicitly subdivide the domain into three zones - solid, liquid, and mushy; nevertheless, these zones could be identified from the computational results. Usually, however, heuristic assumptions should be made to close the system of the governing equations, concerning, for example, the effective thermal conductivity of the solid–liquid mixture.

The whole domain Ω is divided into a solid Ω_s and a liquid Ω_l parts: $\Omega = \Omega_s(t) \cup \Omega_l(t)$ separated by the moving boundary of the crystallization front $\Gamma_0(t)$.

In the phase field methods, a function $f(x)$ is defined as

$$f(x) = \begin{cases} 1, & x \in \Omega_s \\ 0, & x \in \Omega_l \end{cases}$$

with the interface being captured by a value between 0 and 1.

In the level set method, a smooth function $\xi(x, t)$ is introduced instead of $f(x)$ such that

$$\xi(x,t) = \begin{cases} > 0, & x \in \Omega_s \\ = 0, & x \in \Gamma_0 \\ < 0, & x \in \Omega_l \end{cases}$$

and

$$\frac{\partial \xi}{\partial t} + V \cdot \nabla \xi = 0,$$

where V is the interface velocity. The unit normal n_0 to the interface is given as

$$n_0 = -\frac{\nabla \xi}{|\nabla \xi|},$$

with n_0 directing outside of the solid phase and the mean curvature of the interface as

$$K = \nabla \cdot n_0.$$

Variables of the level set method and the phase field method could be related with the use of the Heaviside function.

The interface integral over Γ_0 could be converted into the volume integral

$$\int_{\Gamma_0} (\ldots) d\Gamma = \int_{\Omega} (\ldots) \delta(\xi) |\nabla \xi| d\Omega.$$

Usually a signed distance - the shortest distance from the point x to interface, which is positive inside Ω_s and negative inside Ω_l - is used for the definition of function $\xi(x, t)$. However, this choice will result in the huge computational grids needed to resolve the complex shape of the crystallization front and is practical for the simulation of the growth of the single or a few dendrites.

The detailed description of the crystallization front geometry is not needed in cryobiology problems, where the primary interest is the temperature and solute distributions during freezing that define the cell's behavior. Liu et al. introduced a novel level set function, which is expected to display less fluctuations in the mushy zone. They proposed to use the difference between a "*virtual temperature*" (defined below) and a real temperature instead of the signed distance.

The authors considered coupled heat and mass transfer in the freezing suspension neglecting convective effects – both heat and solutes are transported by molecular diffusion. The authors' remark that "fluid motion is constrained by tissue microstructures" seems, however, to be irrelevant for the considered processes in cells suspension.

Heat conduction equation for the local temperature $\theta(x, t)$ is used for both liquid and solid regions:

$$C_i \frac{\partial \theta_i}{\partial t} + \nabla \cdot \mathrm{q}_i = 0, \qquad ...(5)$$

where C_i is the heat capacity for *i*th phase and the heat flux q_i obeys the Fourier law:

$$\mathrm{q}_i = -\hat{\Lambda}_i \cdot \nabla \theta_i, \qquad ...(6)$$

where $\hat{\Lambda}_i$ is the thermal conductivity tensor. The latent heat of crystallization per unit volume L enters the condition on the freezing front Γ_0 that has the normal n_0 and moves with the velocity V:

$$\|q\| = LV \cdot \mathrm{n}_0, \qquad ...(7)$$

where $\|\cdot\|$ denotes the jump of the variable f across the interface $\|f\| = f_l - f_s$.

Similarly, assuming solute diffusion in the liquid and solid phases:

$$\frac{\partial c_i}{\partial t} + \nabla \cdot \mathrm{j}_i = 0, \qquad ...(8)$$

where the diffusion flux j_i driven by the concentration gradient is

$$\mathrm{j}_i = -\hat{D}_i \cdot \nabla c_i. \qquad ...(9)$$

Here $\hat{D}_i$ is the solute diffusivity tensor. Rejection of the solute from the ice crystallization front is described as

$$\|\mathrm{j}\| \cdot \mathrm{n}_0 = \|c\| \mathrm{V} \cdot \mathrm{n}_0. \qquad ...(10)$$

The authors convert the concentration differences in the solid or liquid solutions into so-called *solidus* or *liquidus* temperatures using phase diagrams that describe dependence of the phase state on the temperature and concentration. Phase diagram was simplified by approximation of the real liquidus and solidus curves by straight lines segments such as

$$\frac{\partial c_i}{\partial T_i} = -\sigma_i, \qquad \text{...(11)}$$

where σ_l and σ_s are the slopes of the liquidus and solidus curves. Then the following inequalities are evidently valid:

$$T_l - \theta < 0 \quad \text{in liquid}, \qquad \text{...(12)}$$

$$T_s - \theta > 0 \quad \text{in solid}. \qquad \text{...(13)}$$

Since $T_l = T_s$ at the interface Γ_0, the liquidus and solidus temperatures could be united into a single variable T named "virtual temperature" and one can write

$$\|c\| = -\|\sigma\| T. \qquad \text{...(14)}$$

The crystallization rate in cryopreservation is not very high; therefore, it is reasonably to disregard "*kinetic supercooling*" (i.e., ignore the possible role of the attachment of water molecules to the ice surface as a rate-limiting stage of the crystal growth) and to use the Gibbs–Thomson relation for the melting temperature depression at the interface:

$$\theta = T - \chi = T - \Gamma K, \qquad \text{...(15)}$$

where χ is the capillary freezing point depression due to the interface curvature K and Γ is a capillary constant, which is $\Gamma \approx 0.025\ K\ \mu$m for the water/ice interface. The Gibbs–Thomson correction to the melting temperature could be ignored for the freezing of cell suspensions in bulk volume, thus $\theta \approx T$. Then the following relations are valid:

$$T - \theta < 0 \quad \text{in} \quad \Omega_l, \qquad \text{...(16)}$$

$$T - \theta = 0 \quad \text{on} \quad \Gamma_0, \qquad \text{...(17)}$$

$$T - \theta > 0 \quad \text{in} \quad \Omega_s. \qquad \text{...(18)}$$

Evidently, the difference between the virtual and real temperature could serve as a level set function:

$$\xi(x,t) = T(x,t) - \theta(x,t).$$

The external boundary of the computational domain with the normal n is divided into patches on which the temperature/concentration or the corresponding heat or mass fluxes are given:

$$\theta = \tilde{\theta} \qquad \text{on} \qquad \Gamma_\theta, \qquad \text{...(19)}$$

$$\mathrm{q}\cdot\mathrm{n}=\tilde{q} \quad \text{on} \quad \Gamma_q, \qquad \text{...(20)}$$

$$c=\tilde{c} \quad \text{on} \quad \Gamma_c, \qquad \text{...(21)}$$

$$\mathrm{j}\cdot\mathrm{n}=\tilde{j} \quad \text{on} \quad \Gamma_j. \qquad \text{...(22)}$$

Conditions at the interface can be written in a weak form as

$$(\Gamma_0,\phi)=\int_{\Gamma_0}\phi\left(\|\mathrm{q}\|\cdot\mathrm{n}_0-L\mathrm{V}\cdot\mathrm{n}_0\right)d\Gamma, \qquad \text{...(23)}$$

$$(\Gamma_0,\psi)=\int_{\Gamma_0}\psi\left(\|\mathrm{j}\|\cdot\mathrm{n}_0-\|c\|\mathrm{V}\cdot\mathrm{n}_0\right)d\Gamma. \qquad \text{...(24)}$$

Here ϕ and ψ are admissible test function for the heat transfer and mass transfer, respectively. Conversion of the last equations into the volume integrals yields

$$(\Gamma_0,\phi)=\int_\Omega\phi\left(-\|\mathrm{q}\|+\|u\|\mathrm{V}\right)\cdot\delta(\xi)\nabla\xi\, d\Omega \qquad \text{...(25)}$$

$$=\int_\Omega\phi\left[\sum_{i=l,s}\nabla f_i\cdot\mathrm{q}_i+\sum_{i=l,s}u_i\frac{\partial f_i}{\partial t}\right]d\Omega, \qquad \text{...(26)}$$

$$(\Gamma_0,\psi)=\int_\Omega\psi\left(-\|\mathrm{j}\|+\|c\|\mathrm{V}\right)\cdot\delta(\xi)\nabla\xi\, d\Omega \qquad \text{...(27)}$$

$$=\int_\Omega\psi\left[\sum_{i=l,s}\nabla f_i\cdot\mathrm{j}_i+\sum_{i=l,s}c_i\frac{\partial f_i}{\partial t}\right]d\Omega. \qquad \text{...(28)}$$

This transformation is an important component of the approach within which the moving boundary problem is treated using a fixed computational grid. Similar transformations are used to convert the volume integrals defined over the liquid and solid subdomains into integral over the entire domain.

A variational formulation for the heat and mass transfer in the system comprising two subdomains - solid and liquid, the interface and the external boundary can be written as

$$\Pi_1\equiv(\Omega,\phi)+(\Gamma_0,\phi)+(B,\phi)=0,$$

$$\Pi_2\equiv(\Omega,\psi)+(\Gamma_0,\psi)+(B,\psi)=0,$$

where (details of the derivation could be found in the cited paper by Liu et al.)

$$(\Omega,\phi)=\int_\Omega\phi\left(\sum_{i=l,s}f_i\frac{\partial u_i}{\partial t}+\sum_{i=l,s}f_i\nabla\cdot q_i\right)d\Omega,$$

$$(B,\phi)=\int_{\Gamma_q}\phi(\tilde{q}-q\cdot n)d\Gamma,$$

$$(\Omega,\psi) = \int_{\Omega}\psi\left(\sum_{i=l,s} f_i\frac{\partial u_i}{\partial t} + \sum_{i=l,s} f_i\nabla\cdot j_i\right)d\Omega,$$

$$(B,\psi) = \int_{\Gamma_q}\psi(\tilde{j} - j\cdot n)d\Gamma.$$

Using the continuity of the temperature θ in both phases, one can write

$$\Pi_1 = \int_{\Omega}\left[\frac{\partial\tilde{u}}{\partial t} - \nabla\cdot(\tilde{k}\cdot\nabla\theta\right]d\Omega + (B,\phi) = 0, \quad ...(29)$$

where for brevity the following notation was introduced:

$$\tilde{u} = \sum_{i=l,s} f_i u_i \quad \text{and} \quad \tilde{k} = \sum_{i=l,s} f_i k_i.$$

This procedure could not be used directly for mass transfer since the concentration is not continuous at the interface. However, the virtual temperature is and thus could be used instead to get

$$\Pi_2 = \int_{\Omega}\left[\frac{\partial\tilde{c}}{\partial t} - \nabla\cdot(\tilde{D}\cdot\nabla T)\right]d\Omega + (B,\psi) = 0. \quad ...(30)$$

where

$$\tilde{c} = \sum_{i=l,s} f_i c_i \quad \text{and} \quad \tilde{D} = \sum_{i=l,s} f_i D_i.$$

Liu et al. used Galerkin finite elements method, that is, set $\phi = \theta$, $\psi = T$, getting the coupled equations:

$$\int_{\Omega}\theta\frac{\partial\tilde{u}}{\partial t}d\Omega + \int_{\Omega}\nabla\theta\cdot\tilde{k}\cdot\nabla\theta d\Omega = -\int_{\Gamma_q}\theta\tilde{q}d\Gamma, \quad ...(31)$$

$$\int_{\Omega}T\frac{\partial\tilde{c}}{\partial t}d\Omega + \int_{\Omega}\nabla T\cdot\tilde{D}\cdot\nabla T\Omega = -\int_{\Gamma_i}T\tilde{j}\,d\Gamma, \quad ...(32)$$

that contain only the characteristic functions f_i and does not depend on the particular choice of the level set function $\xi(x, t)$.

Isoparametric quadrilateral finite elements with linear shape functions were used for the spatial approximation. To solve the discrete equations, the authors used a first order in time Euler backward differences and a modified Newton iterative procedure to deal with nonlinearity. The Heaviside $H(x)$ and delta $\delta(x)$ functions present in the problem formulation were approximated as

$$H_{\varepsilon}(x) = \begin{cases} 1, & x > \varepsilon \\ 0, & x > \varepsilon \\ \frac{1}{2}\left[1 + \frac{x}{\varepsilon} + \frac{1}{\pi}\sin\left(\frac{\pi x}{\varepsilon}\right)\right], & |x| \le \varepsilon \end{cases}$$

$$\delta_\varepsilon(x) = \begin{cases} 0, & |x| > \varepsilon \\ \frac{1}{2\varepsilon}\left[1+\cos\left(\frac{\pi x}{\varepsilon}\right)\right], & |x| \le \varepsilon \end{cases}$$

The computational results were validated using data obtained by Korber et al. in the experiments on planar freezing of sodium permanganate solutions and applied to the freezing of suspension of AT-1 Dunning rat tumor cells. The results showed, in particular, that the solute concentration could be as large as six times the original isotonic concentration.

Explicit Treatment of "Mushy" Zone

Chua et al. in their study of effects of cryosurgery on selective cell destruction adopted an approach in which the mushy zone is treated explicitly as one of subdomains of the whole computational domain. The authors used special forms of the Pennes bioheat equation in the different domains:

1. In the frozen region, taking into account the absence of the blood perfusion and zero metabolic rate

$$\rho_s c_s \frac{\partial T_s(\mathrm{x},t)}{\partial t} = \nabla k_s \nabla T_s(\mathrm{x},t),$$

2. In the mushy region

$$\rho_i c_i \frac{\partial T_i(\mathrm{x},t)}{\partial t} = \nabla k_i \nabla T_i(\mathrm{x},t) + \Omega_b \rho_b c_b (T_b - T_i(\mathrm{x},t) + Q_m + \rho_i h_L \frac{df_s}{dt},$$

3. In the unfrozen region

$$\rho_L c_L \frac{\partial T_L(\mathrm{x},t)}{\partial t} = \nabla k_L \nabla T_L(\mathrm{x},t) + \Omega_b \rho_b c_b (T_b - T_L(\mathrm{x},t)) + Q_m,$$

where the solid fraction during the phase change f_s had been introduced, which was defined using the temperatures of liquidus and solidus (the authors have assumed the latter to be constant and equal to 272 and 265 K, respectively, and to define the boundaries of the mushy zone):

$$f_s = \begin{cases} 0, & T_i(\mathrm{x},t) = T_{\text{liquidus}} \\ 1, & T_i(\mathrm{x},t) = T_{\text{solidus}} \\ \frac{T_i(\mathrm{x},t) - T_{\text{liquidus}}}{T_{\text{solidus}} - T_{\text{liquidus}}}, & T_{\text{solidus}} < T_i(\mathrm{x},t) < T_{\text{liquidus}} \end{cases}$$

The equations describing heat transfer in these three domains - unfrozen, mushy, and frozen - were coupled via the boundary condition that express the continuity of the temperature and heat fluxes at the interfaces between the domains:

1. At the boundary between the frozen and mushy domains:

$$T_s(\mathrm{x},t) = T_i(\mathrm{x},t) = T_{\text{solidus}}, \qquad \text{...(33)}$$

$$\lambda_s \nabla T_s(\mathrm{x},t)\cdot\hat{n} = \lambda_i \nabla T_i(\mathrm{x},t)\cdot\hat{n} \qquad \text{...(34)}$$

2. At the boundary between the mushy and unfrozen domains

$$T_i(\mathrm{x},t) = T_L(\mathrm{x},t) = T_{\text{liquidus}}, \qquad \text{...(35)}$$

$$\lambda_i \nabla T_i(\mathrm{x},t)\cdot\hat{n} = \lambda_L \nabla T_L(\mathrm{x},t)\cdot\hat{n} \qquad \text{...(36)}$$

Similar boundary conditions are specified for the temperature at the interface between unfrozen domain and surrounding tissue (subscript "ST" is used below) while no heat exchange (adiabatic thermal condition) is allowed through this boundary:

$$T_L(\mathrm{x},t) = T_{ST}(\mathrm{x},t), \qquad \text{...(37)}$$

$$\lambda_L \nabla T_L(\mathrm{x},t)\cdot\hat{n} = 0. \qquad \text{...(38)}$$

For the integration of equations in time, Chua et al. used two-level finite difference scheme

$$\int_t^{t+\Delta t} T(\mathrm{x},t)dt = [wT(\mathrm{x},t+\Delta t) + (1-w)T(\mathrm{x},t)]\Delta t,$$

with a weighting factor $w \in (0, 1)$ (the choice $w = 0.5$ corresponds to the well known Crank–Nicolson scheme that is second order accurate in time). Spatial discretization is accomplished using the finite volume method.

After validating the numerical model using the experimental data by Seifert et al., the authors considered several spherical tumor regions different in diameter (chosen in such a way as to represent different stages of the tumor development). They also discussed the case of the forced thawing and note that in this case two frozen/unfrozen interfaces are present in the system, the first being the primary crystallization front that is recessing due to the switch-off of the heat removal mechanism and the second one is formed at the probe surface, which is heating at this stage and propagates away from the probe. Chua et al. did not consider the interaction of these interfaces when they meet.

The authors studied the effect of the cryoprobe diameter (and, hence, the rate of the heat removal from the tissues, since the surface temperature is essentially independent of the probe diameter) and efficiency of the repeated freezing; the latter is found to increase the cell destruction at least as much as 20%.

Cell Interaction with Ice Front

As was discussed above, the behavior of the cell in respect to the moving ice crystallization front depends on the velocity of the latter:

the cell are forced from the phase boundary if the front velocity exceeds some threshold value V_c, and are trapped by the ice for smaller velocities. There is a balance between two forces acting on the cell – thermomolecular force due to the cell interaction with the ice surface and the viscous force due to the fluid flow. There is, however, another force that arises when the concentration of solute changes considerably on the length scale of the order of the cell diameter. Solute gradient in the freezing suspension is formed near the ice crystallization front due to the solute rejection from the growing ice crystal. Cell migration caused by this force is considered below; the simulation of the process of cell trapping in the growing ice crystal is discussed later.

Migration of Cells in the Solute Concentration Gradient

When a cell with a permeable membrane is placed in the solution with the solute concentration gradient, the net force acting on the cell is not zero, since the water flux across the cellular membrane does depend on the local solute concentration, which is unequal on the opposite cell's sides. The situation for the case of constant gradient where one-dimensional solute concentration is shown along the line through the cell's centre x_c. This distribution is plotted under the usual assumption that the cytoplasm is well mixed and thus solution concentration inside the cell is constant. It is assumed that solute could not move across the cellular membrane and the overall cell's motion is defined by the water flux through the membrane. Evidently, water will leave the cell on the side facing the higher solute concentration (hypertonic environment) and will flow into the cell on the side facing the lower concentration (hypotonic environment). The process will be accompanied by the cell volume excursions (either shrinking or swelling), if the average solute concentration differs from the concentration inside the cell.

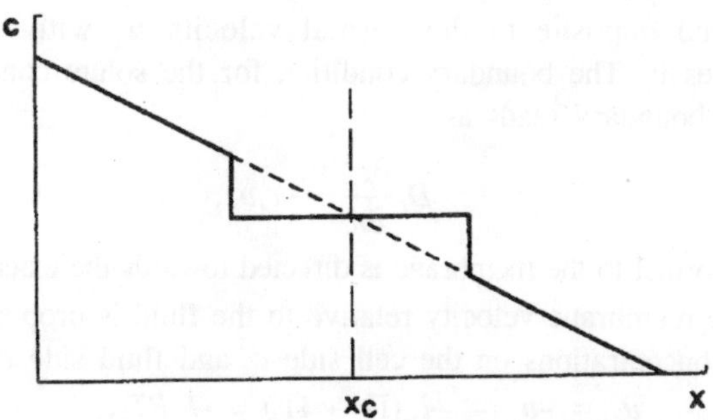

Fig. 8.1. Solute concentration distribution in and around the cell in the constant gradient.

This problem was considered both analytically for the spherical and cylindrical cells in the solution with the constant solute concentration and numerically by Jaeger et al. The authors named the resulting cell motion *osmotic migration* to acknowledge the diving forces of the process. Jaeger et al. assumed that velocity of the cellular membrane is small and the variation of the cell's volume can be considered as quasi stationary one. To avoid complications related to the possible alteration of the average density of the cell and thus buoyancy effects, the concentration gradient is considered to be perpendicular to the direction of the gravity and the temperature is taken to be constant. The effect of the cellular organelles on the osmotic response of the cell is accounting for by explicit incorporation of the osmotically inactive volume. Effects of the concentration dependence of the interfacial energy are disregarded. The concrete computations were performed for the binary solution, with parameters equal to those of aqeous NaCl solution.

Under the assumption of no bulk fluid motion, the solute concentration is governed by the diffusion equation written earlier as the system of two first order equations (8) and (9) for the concentration and the diffusion flux. It is convenient to rewrite it as a second order parabolic equation for the concentration as

$$\frac{\partial c_i}{\partial t} = \nabla \cdot D_i \nabla c_i, \qquad \text{...(39)}$$

where the subscript "*i*" refers to either the cell ($i = c$) or the surrounding fluid ($i = f$) and diffusion is assumed to be isotropic and thus the diffusivity tensor is reduced to the scalar diffusion coefficient D_i.

The impermeable to the solute membrane move with velocity u_n equal and opposite to the normal velocity u_w with which water permeates it. The boundary condition for the solute concentration at the cell boundary reads as

$$D_i \frac{\partial c_i}{\partial n} = -c_i u_n, \qquad \text{...(40)}$$

where normal to the membrane is directed towards the external medium.

The membrane velocity relative to the fluid is proportional to the solute concentrations on the cell side c_c and fluid side c_f:

$$u_n = -u_w = -L_p(\Pi_f - \Pi_c) \approx -L_p RT(c_f - c_c), \qquad \text{...(41)}$$

where L_p is the hydraulic permeability and Π_i is the osmotic pressure in the medium *i*.

In the case of the uniform solution, (41) describes cell shrinking ($u_n < 0$) in the hypertonic environment and cell swelling ($u_n > 0$) in the hypotonic one.

Assuming length and concentration scales to be given by the initial cell's radius r_0 and the concentration difference Δc, the authors get the velocity scale as $U = LRT\Delta c$ and the Peclet number as

$$Pe = \frac{r_0 U}{D_i}.$$

Spherical and a cylindrical cell in a uniform concentration gradient

For the case of one spherical cell in an infinite medium and the constant diffusion coefficient, (39) reduces to the Laplace equation with zero normal gradient at the cell surface. Classical solution (obtained by Lamb) is well known:

$$c_f = c_f^0 - Gx_c - \frac{3}{2}Gr\cos\theta,$$

where r is the radius of the cell, x_c is the position of the cell centroid; c_f^0 is the concentration in the fluid at $x = 0$, and G is the solute gradient far from the cell. According to (41), the normal velocity of the cellular membrane is then

$$u_n(\theta) = -L_p RT(c_f^0 - Gx_c - \frac{3}{2}Gr\cos\theta - c_c)$$

and the change of volume of the cell

$$\frac{dV}{dt} = \oint_S u_n ds = -L_p RT \int_0^{2\pi}\int_0^{\pi}\left(c_f^0 - Gx_c - \frac{3}{2}Gr\cos\theta - c_c\right)r^2 \sin\theta\, d\theta d\varphi$$
$$= 4\pi L_p RTr^2(c_f^0 - Gx_c - c_c).$$

The last equation easily yields the relation for the change of cell radius in time, exploiting the constancy of the solute amount in the cell.

The velocity of the cell's centroid is calculated as a variation of the integral over the cell volume V:

$$\frac{dx_c}{dt} = \frac{d}{dt}\left(\frac{1}{V}\int_V x dv\right) = \frac{1}{V}\frac{d}{dt}\int_V x dv - \frac{x_c}{V}\frac{dV}{dt},$$

with the use of the relation

$$\frac{d}{dt}\int_V x dv = \oint_S x u_n ds.$$

The case of a cylindrical cell is treated similarly. The final expressions for the evolution of the cell radius and the centroid velocity for these two cell types obtained by Jaeger et al.

The equations for the cell radius and the centroid velocities could be written in dimensionless form using the above defined velocity scale U, initial cell radius r_0, time scale $1/L_pRTG$ and a product of the fluid concentration gradient, and r_0 as the concentration scale $\Delta c = Gr_0$ to get for the spherical cell:

$$\frac{d\tilde{r}}{d\tilde{t}} = -\left(\tilde{c}_f^0 - \tilde{x}_c - \tilde{c}_c^0\left(\frac{1}{\tilde{r}}\right)^3\right), \quad ...(42)$$

$$\frac{d\tilde{x}_c}{d\tilde{t}} = \frac{3}{2}\tilde{r}, \quad ...(43)$$

and similarily for the cylindrical one:

$$\frac{d\tilde{r}}{d\tilde{t}} = -\left(\tilde{c}_f^0 - \tilde{x}_c - \tilde{c}_c^0\left(\frac{1}{\tilde{r}}\right)^2\right), \quad ...(44)$$

$$\frac{d\tilde{x}_c}{d\tilde{t}} = 2\tilde{r}. \quad ...(45)$$

When the initial cell concentration is zero, the analytical solution could be written out as (for the spherical case)

$$x(t) = \tilde{c}_f^0\left(1 - \cosh\sqrt{\frac{3}{2}}t\right) + \sqrt{\frac{3}{2}}\sinh\sqrt{\frac{3}{2}}t,$$

$$r(t) = -\sqrt{\frac{3}{2}}\tilde{c}_f^0\sinh\sqrt{\frac{3}{2}}t + \cosh\sqrt{\frac{3}{2}}t.$$

The cell shrinks to the zero radius after a finite time. The dependence of solution on both the fluid concentration and the initial cell concentration rather that on just the concentration difference explained by the authors by consideration of the limiting case of zero initial cell concentration (that will remain zero independent of the volume excursions), while for the nonzero c_c^0 cell concentration will follow the volume changes (the total solute content in the cell is fixed).

If the nonzero osmotically inactive volume V_b should be taken into account, the cell concentration is written as

$$c_c = c_c^0\frac{V_0 - V_b}{V - V_b}$$

and the evolution of the cell radius will be governed by

$$\frac{dr}{dt} = -L_p RT\left(c_f^0 - Gx_c - c_c^0\left(\frac{1-\alpha}{\left(\frac{r}{r_0}\right)^3 - \alpha}\right)\right),$$

while the centroid velocity being independent of the cell concentration will remain the same as in the case when the osmotically inactive volume can be disregarded; here $\alpha = \upsilon_b/V_0$. Computations performed by Yaeger et al. for the erythrocytes showed that account for the osmotically inactive volume yields a difference in the centroid velocity of less than 5%.

The general pattern of the cell behavior depends on the environment solute concentration relative to that inside the cell. In the isotonic environment ($c_f^0 = c_c^0$), the cell moves towards a lower environment solute concentration, increasing its size and the centroid velocity since the latter is proportional to the radius. In case of the hypertonic ($c_f^0 > c_c^0$) or hypotonic ($c_f^0 < c_c^0$) environment two process stage could be distinguished: (1) rapid shrinking or swelling of the cell to reach the concentration level intermediate to the concentrations in fluid on two cell sides and (2) migration towards the lower concentration as in the isotonic case. The authors' estimates give the typical centroid velocity for erythrocytes ca. 0.2 μm s^{-1}, with 2D approximation overestimating 3D case by about 30%.

The general case of nonconstant fluid concentration gradient was considered in the two-dimensional approximation for the typical concentration behavior near the moving ice crystallization front (the concentration gradient depends on the front velocity, which in turn is determined by the cooling rate and reduces with the distance from the front) using finite element computations.

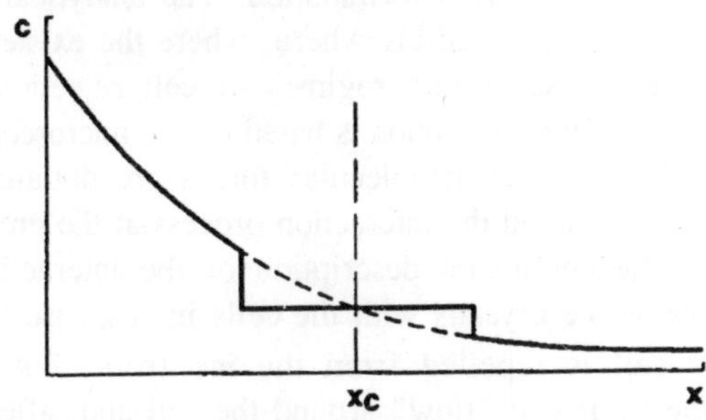

Fig. 8.2. Solute concentration distribution in and around the cell near the ice crystallization front.

FEM computations have also been used to simulate the behavior of several identical cells lined up parallel to the ice front and show that in this case the migration velocity increases with the decrease of the gap between the cells. This effect is explained by the blockage of the solute diffusion from the crystallization front and greater drop of the concentration across the cell diameter.

To explain the observed experimental alignment of the cells in front of the solidification interface, the authors performed computations for the perturbed positions of the cells and found that osmotic migration could provide the perfect cell alignment. That really could be the reason, but one comment is in order. As the authors themselves noted, the forces involved in the osmotic migration are rather small as well as the resulting centroid velocities (about 12% of the moving ice front velocity, according to the authors' estimates) and could not be responsible for the motion of the cells from the front. It is unclear whether it is worth to draw this rather weak osmotic migration for the explanation of phenomena that could be easily produced by already mentioned thermomolecular force of the cell interaction with the ice surface. This force, being depending inversely on the cube of the distance between the cell and the ice front (and being balanced by the viscous force of the fluid flow), could, evidently, provide the mechanism for the cells' alignment near the front.

Another notion by Jaeger et al. is of greater interest. The authors note that in spite of the cell velocity being small compared to the front velocity, it could be measured and used for experimental determination of cell properties such as permeability.

Cell Interaction with the Advancing Plane Crystallization Front

The interaction of a single biological cell with the moving ice crystallization front was considered by Carin and Jaeger using an arbitrary Lagrangian- Eulerian formulation. The analytical treatment of this problem was considered elsewhere, where the existence of the critical front velocity separating regimes of cell rejection and cell trapping was shown. This description is based on the microscopic picture of the process and the thermomolecular forces are dominant. Carin and Jaeger have considered the interaction process at the greater space scale, allowing the continuum description of the interaction of the finger-like growing ice crystals with the cells in suspense.

When the cell is repelled from the ice front, but slows its propagation, the front will "flow" around the cell and, after merging of the two front branches, the cell will be engulfed by the ice crystal.

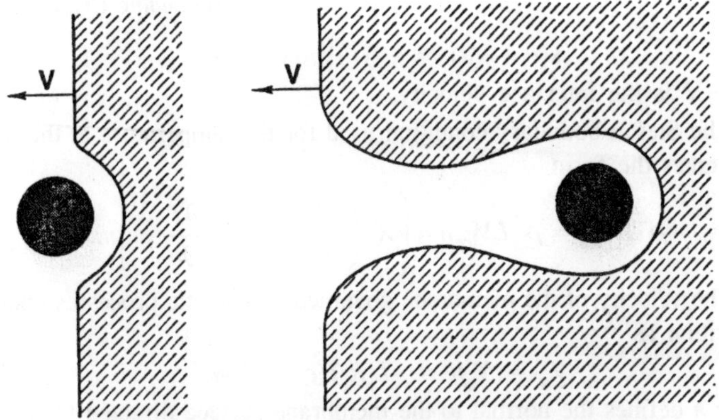

Fig. 8.3. Trapping of the cell by the advancing ice front.

Carin and Jaeger considered 2D (plane or axisymmetric) problem of the cell behavior in the aqueous solution of salt (NaCl) near the flat or destabilized ice crystallization front under several assumptions, including the following ones:

1. The intracellular and extracellular media are binary solutions with parameters of the aqeous NaCl solution
2. Solute is completely rejected by the ice front
3. Solute is nonpermeating
4. The thermophysical properties of the ice, the extracellular, and the intracellular fluids are constant
5. The mechanical action of membrane is disregarded
6. No fluid motion is induced by the membrane
7. No gravity

Three phases were considered: one or more cells (Ω_c), the extracellular solution (Ω_l), and ice (Ω_s). The heat transfer was described assuming that the only mechanism is conduction (5) and (6); this system was rewritten as a second order equation for the temperature:

$$\rho c_p \frac{\partial T}{\partial t} = \nabla \cdot (\lambda \nabla T) \qquad ...(46)$$

that is solved in all three domains Ω_c, Ω_l, and Ω_s.

The concentration field evolution was described by which is solved in the domains Ω_c and Ω_l.

The different phases are separated by the interfaces that are either cell membranes ($\Gamma_{cl} = \partial\Omega_c = \partial\Omega_c \cap \partial\Omega_l$) or the crystallization front ($\Gamma_{sl} = \Omega_l \cap \Omega_s$). The position of each interface is determined using the first order approximation in time as

$$x(t + \Delta t) = x(t) + \Delta t V(x, t),$$

where V is the interface velocity.

The Stefan boundary condition used for the temperature at the ice front takes the form

$$\rho^- L V_{ice} = \rho^- L V_{ice} n = \left(\lambda^- \left(\frac{\partial T}{\partial n} \right)^- - \lambda^+ \left(\frac{\partial T}{\partial n} \right)^+ \right) n. \quad ...(47)$$

The velocity of the membrane governed by osmotic processes could be written as

$$V_m = V_m n = -L_p R_g T(c^+ - c^-)n, \quad ...(48)$$

where n defines the normal to the membrane surface directed towards the external medium, R_g is the universal gas constant. The temperature dependence of the hydraulic permeability is given by.

Evidently, in the homogeneous external medium, (48) describes cell shrinking in the hypertonic environment ($V_m < 0$) and cell swelling ($V_m > 0$) in the hypotonic one.

Rejection of the solute by both the ice crystallization front and by the cellular membrane is formulated as

$$D^{\pm} \nabla c^{\pm} n = D^{\pm} \left(\frac{\partial c}{\partial n} \right)^+ = -c^{\pm} V n,$$

where V is the velocity of the interface and the superscript $\pm$ denotes the sides of the interface.

The deviation of the temperature at the ice front is due to the Gibbs- Thomson effect written as

$$T = T_{m0} + mc^+ - \frac{\Gamma T_{m0}}{\rho L} K,$$

where T_{m0} is the equilibrium melting temperature, m is the slope of the linearized liquidus line from the equilibrium phase diagram, c^+ is the solute concentration in the liquid phase, Γ is the surface tension, and K is the curvature of the ice front.The negative sign of m as for the salt indicate the depressing of the melting point with increase of the salt concentration.

The method used by Carin and Jaeger is constructed by the combination of the two concepts – the arbitrary Lagrangian–Eulerian (ALE) description exploiting moving grids whose velocity in general case differs from the material velocity, and front tracking, thus the authors named it front tracking ALE (FTALE). Implementation of this approach involves two separate grids (meshes) – domain mesh and interface mesh. The three-node triangular elements are used for the former; according to its position in respect to the interface, three types of the elements in the domain mesh are distinguished:

1. Element that completely belongs to one domain
2. Element that completely belongs to one domain but one or two of its nodes are located at the interface
3. Element that is crossed by the interface.

Equations (46) and (39) are transformed to account for the mesh movement by introduction of the extra advection term in each of these equations as

$$\rho c_p\left(\frac{\partial T}{\partial t}-V^{\text{ALE}}\cdot\nabla T\right)=\nabla\cdot(\lambda\nabla t), \qquad \text{...(49)}$$

$$\frac{\partial c_i}{\partial t}-V^{\text{ALE}}\cdot\nabla c=\nabla\cdot D_i\nabla c_i. \qquad \text{...(50)}$$

The integral forms I_t and I_c associated with the weak formulation of the heat and mass transfer problems are:

$$I_t=\int_{\Omega}\left[\rho c_p\delta T\left(\frac{\partial T}{\partial t}-V^{\text{ALE}}\cdot\nabla T\right)+\nabla\delta T\cdot\nabla T\right]dS$$

$$-\int_{\partial\Omega}\lambda\delta T\frac{\partial T}{\partial n}dl-\int_{\Gamma_{\text{ice}}}\rho^{-}L\delta TV_{\text{ice}}dl, \qquad \text{...(51)}$$

$$I_c=\int_{\Omega}\left[\delta c\left(\frac{\partial c}{\partial t}-V^{\text{ALE}}\cdot\nabla c\right)+\nabla\delta c\cdot\nabla c\right]dS-\int_{\partial\Omega}\lambda\delta c\frac{\partial c}{\partial n}dl$$

$$-\int_{\Gamma_{\text{ice}}}\delta cc^{+}V_{\text{ice}}dl-\int_{\Gamma_{\text{m}}}\delta c^{+}c^{+}V_m dl+\int_{\Gamma_{\text{m}}}\delta c^{-}c^{-}V_m dl, \qquad \text{...(52)}$$

where δT and δc are weighting functions.

The Stefan condition (47) is used to determine the velocity of the ice crystallization front. When linear triangular elements are used for the spatial approximation, the first derivatives are constant over the element. Since the nodal values of the derivatives are needed, some reconstruction procedure must be invoked. The authors have found that this approach could result in the instability on the nonuniform meshes and thus suggested an alternative – to use the weak form of the Stefan condition (47) that reads as

$$\int_{\Gamma_{\text{ice}}}\rho^{-}L\delta V_{\text{ice}}V_{\text{ice}}dl=\int_{\Gamma_{\text{ice}}}\delta V_{\text{ice}}\left(\lambda^{-}\left(\frac{\partial T}{\partial n}\right)^{-}-\lambda^{+}\left(\frac{\partial T}{\partial n}\right)^{+}\right)dl.$$

For the verification of the proposed approach the exact solution of the Stefan problem in cylindrical coordinates obtained by Paterson is used. The growth of the cylindrical solid from the supercooled pure melt was computed using 2D Cartesian coordinates.

As was noted in previously, solution of the heat and mass transfer in the freezing suspension and computation of the forces acting on the cells due to nonuniformity of the environment and nonsymmetric character of the water fluxes into/out of the cell could not account for the cell's repulsion from the advancing ice crystal front. The latter is provided by the thermomolecular forces disregarded in the earlier described problem statement. Thus the authors consider the case of the front speed much greater than the critical velocity for the entrapment. The presence of the cell in the liquid phase forces the ice crystal to become concave. The computations performed by the authors for the red blood cells idealized as spheres showed that cells present a significant obstruction for the solute diffusion.

To study the effect of the permeability of the cellular membrane, a comparison was made between the ice front interaction with the cell and the rigid solid particle, whose properties except permeability are the same as properties of the cell. The ice front was found to advance more slowly in the case of particle that was attributed to size difference due to the cell shrinking; cell water loss by exosmosis was found to be too small to have a noticeable effect on the solute distribution. It is anticipated that exosmosis effect will be greater for the lower cooling rate, since cell will then have more time for the equilibration of the cytosol composition with the environment solution.

To simulate the cell interaction with the destabilized crystallization front, the authors forced a sinusoidal perturbation of the moving front and found this perturbation to be dumped in the course of cell engulfment. It was also observed that the width of the forming channel increases with increase of the initial distance of the cell to the crystallization front that is explained by the greater solute concentration in the latter case accumulated due to the decrease of the volume of the liquid phase (at the constant amount of the solute) in the course of crystallization front propagation. The same processes are responsible for the delay in the cell shrinking that increases with the initial distance to the ice front. The authors note that this nonuniformity in the osmotic response usually disregarded in the analysis of the freezing of cell suspense could be significant in some cases.

Cell in the Freezing Cylindrical Cavity

Mao et al. also considered the cell interaction with advancing ice front, but in somewhat different context compared with the problem discussed in previous section. They analyze the cell behavior for the cylindrical cell (axisymmetric problem is considered) that initially resides in the center of salt solution confined by the cylindrical ice front that is concentric with the cell. The difference from the problem considered by Carin and Jaeger is not the other geometrical arrangement only, but also the bounded space confinement of the solution and thus anticipated greater variations of its concentration in the course of the crystallization front advancement. Initial size of the cavity is the fourfold cell radius.

The mass transfer (39) is considered. The solute diffusivity dependence on the concentration is neglected. The cell is modeled as an aqueous salt solution enclosed within a semi-permeable thin membrane. The external crystallization front is considered for both stable and unstable growth regimes. In some cases the completely coupled heat (46) and mass transfer was simulated.

The problem is solved using the method for sharp interface tracking on fixed grids developed by Udaykumar et al. and earlier used by the authors to simulate the dendritic solidification of solutions.

An empirical phase diagrams is used; the capillarity effects on the interface temperature were taken in the following form:

$$T_1 = b_0 + b_1 c + b_2 c^2 + b_3 c^3 + b_4 c^4 - \frac{\Gamma_{sl}(\theta)}{L} T_m K,$$

where c is the salt concentration, K is the curvature of the interface, θ is the angle that the normal to the interface makes with the horizontal, and Γ_{sl} is the (anisotropic) surface tension.

The extracellular solution is supposed to be in equilibrium with a planar ice crystallization front, and the salt concentration inside the cell was determined as $c = c_0 V_0 / V$, where the current cell volume V is computed from the ODE:

$$\frac{dV}{dt} = -SJ_w,$$

using the fourth order Runge–Kutta scheme.

The cell shape is assumed to remain circular and the heat transfer through the system to be instantaneous.

Rather large (for 2D problems) computational grids containing up to 500 × 500 cells were used; the insufficient resolution of the 250 ×

250 cells grid was detected by the asymmetrical breakdown of the crystallization front in the course of the instability development.

The authors found that for the permeabilities and cooling rates considered, the cell dehydration was membrane-limited; the rate of the water loss, due to the nonlinearity of the governing equations, was found to be significantly different in the nonisothermal case. Instabilities of the crystallization front have only a moderate effect on the cell response.

Heat Transfer in Biological Tissues

The numerical methods for heat transfer with the phase change are described in numerous papers.

Heat Transfer in Living Tissues

Heat transfer in higher living organisms is characterized by two unique mechanisms: metabolism and blood flow. While the former usually represented as

$$\dot{q}_{met} = \dot{q}_{met0} Q_{10}^{\frac{T_0 - T}{10}},$$

where $T_0 = 37\,°C$ and Q_{10} is a temperature-dependent coefficient frequently taken as constant from the range is of minor importance for cryobiology, the latter should be taken into account.

The difficulties of the mathematical description of the heat transfer in the living tissues originate from both the complexity of the blood vessels' net that could form the fractal structures and changes of the perfusion intensity due to the organism state (e.g., the blood perfusion rate in the active muscle tissue over an order of magnitude larger than that in the muscle at rest) and to the tissues' damage. Frequently arteries and veins are arranged in the so-called counter-current pairs, with the distance between the vessels being about their diameter or less.

The generation of the realistic vascular net is a separate problem; one of the most promising approaches to its solution is the constructive constrained optimization. This methods yields a binary branching network of segments of rigid cylindrical tubes that is contained in a three-dimensional convex volume (perfusion volume) and subjected to a number of the physiological conditions and constraints such as, for example, obeying the bifurcation law (the relation connecting the radiuses of the parent and daughter segments of the form $r_{\text{parent}}^{\gamma} = r_{\text{left}}^{\gamma} + r_{\text{right}}^{\gamma}$).

In addition to the blood flow, there is another process, while certainly of minor importance, that could cause the deviation from the simple conduction heat transfer - the flow of the interstitial liquid occurring due to the temperature nonuniformity, registered in experiments performed by Davydov et al. on the heat transfer in the cow haunch muscle tissue. The sample was cut out in such a way that the muscle fibers were parallel to the surface. Threefold difference in the thermal diffusivities in the direction parallel and perpendicular to the surface was registered. This apparent anisotropy as well as concern about the validity of the common view that the cellular tissue is a passive isotropic media is removed by accounting for heat transfer by the flow of the interstitial fluid.

The models of the blood flow contribution to the global heat transfer could be divided into two groups: continuum and vascular. The construction of the former essentially reduces to the introduction of the efficient thermal conductivity and/or to the addition of the source term to the energy equation.

Continuum Models

The continuum models of the microvascular heat transfer are derived with intention to average the effects of a large number of blood vessels present in the region of interest. The best known and certainly the most important continuum model was suggested by Pennes in 1948 and referred to as the Pennes equation (sometimes this model is also called "heat sink model")

$$\rho c \frac{\partial T}{\partial t} = \nabla \cdot \nabla T + c_b \Omega (T_a - T) + \dot{q}_{\text{met}} + Q^{\text{ext}}, \qquad \text{...(53)}$$

where T, ρ, c, and λ are the temperature, the density, the specific heat, and the thermal conductivity of the tissue as the homogeneous media, Ω is the perfusion coefficient that has the dimensionality of the mass blood flow per unit volume, c_b is the blood specific heat, T_a is the temperature of the arterial blood, $\dot{q}_{\text{met}}$ and Q^{ext} are the heat sources due to the metabolic reactions (this one, as was mentioned, usually could be neglected in the cryobiology problems) and the external source of energy. Pennes formulated his equation studying heat transfer in a human forearm and, most probably, being inspired by studies on the oxygen transport by blood to the cells. The effect of the blood flow in Pennes equation is reduced to the scalar local heat source; directionality of the heat transfer by the blood flow is ignored in this model.

It is assumed that the heat transfer occurs in the capillars only where the blood with temperature T_a flows. In other words, the length of establishing thermal equilibrium is assumed to be infinitely large for all the vessels except capillars and zero for the latters. The temperature in the Pennes model is defined as an average over the volume δV:

$$T(r) = \frac{1}{\delta V} \int_{\delta V} T(r) dV,$$

where the linear scale $3\sqrt{\delta V}$ is assumed to be the following:

1. Small compared with the size of the domain considered while
2. Large compared with the diameter of the "thermally important" vessels

As was noted, no such scale exists for the developed blood vascular net.

Numerous studies show that heat transfer mainly occurs in the greater vessels and the blood flowing into the capillars is already in thermal equilibrium with the surrounding tissues. Still, in practical problems a simple Pennes model frequently gives a more reasonable predictions than some of the more sophisticated models; its combinations with different models based on the effective thermal conductivity are also considered.

The model accounting for the effect of the blood vessel's size on the heat transfer in the tissues introduces the "thermal equilibration length," or simply "thermalization length" L_e as the distance over which the difference between the temperatures of the blood and the surrounding tissue decreases "e" times:

$$\rho c \frac{\partial T}{\partial t} = \nabla \cdot \lambda \nabla T + (\rho c)_b \Omega^* (T_a^* - T) - (\rho c)_b \bar{u} \cdot \nabla T + \nabla \cdot \lambda_p \nabla T + \dot{q}_{\text{met}} + Q^{\text{ext}}. \qquad \text{...(54)}$$

Here the term $(\rho c)_b \bar{u} \cdot \nabla T$ describes the contribution to the heat transfer of the blood flow that is in equilibrium with the tissue. The temperature (as well as the density and specific heat) in this equation should be understood as the volume-weighted parameters for the composite continuum media comprising both the tissue and the blood.

The term $\nabla \cdot \lambda_p \nabla T$ accounts for the additional "perfusional" thermal conductivity:

$$\lambda_p = n(\rho c)_b \pi r_b^2 \bar{V} \cos^2 \Gamma \sum_{i=1}^{\infty} \frac{L_e}{L_e^2 \beta_i^2 + 1},$$

where Γ is the angle between the direction of the blood vessels and the tissue temperature, n is the number of vessels, and r_b is their radius; the parameter β is assumed to be proportional to the vessel length.

The structure of (54) resembles that of the Pennes equation (53); to strengthen the similarity one can unite terms with the temperature gradient.

The thermal equilibration length was discussed by Kotte et al. in the context of their discrete vessels model and an approximate solution for the cylindrical vessels and not too small average blood velocity (allowing to neglect the heat diffusion along the vessel) was derived. The notion of the thermalization length allows one to classify all blood vessels into three groups: large vessels that should be, as a rule, treated explicitly, thermally significant, and thermally insignificant vessels.

The elements of the structure of the blood vessels net (in particularity, the presence of pairs "artery–vein") are incorporated into the three-temperature model of Weinbaum et al. the temperature of the vein blood is considered explicitly in this model. Later the model was simplified down to the single equation:

$$\rho c \frac{\partial T}{\partial t} = \nabla \cdot \lambda_{\text{eff}} \nabla T + \dot{q}_{\text{met}}.$$

The tensor of the effective thermal conductivity is defined as

$$\lambda_{\text{eff}}^{ij} = k_t \left(\delta_{ij} + \frac{n\pi^2 r_b^2}{4\sigma} \left(\frac{\lambda_{b1}}{\lambda_t} \right)^2 Pe^2 l_i l_j \right),$$

where n is the geometric parameter defining the density of the blood vessels, r_b is assumed to be the same for arteries and veins l_i are the direction cosines of the vessels, the Pecle number is $\text{Pe} = 2\rho_{b1} c_{b1} r_b u / \lambda_{b1}$, σ is the parameter that characterizes the heat transfer between the artery and the vein. According to the estimates by the authors of, the anisotropic corrections to the thermal conductivity coeffcient are essential for the blood vessels with diameter greater than 100 μm.

Vascular Models

Numerous vascular models differ greatly by their complexity [933]. There are models based on the concrete anatomical information. Finally, it is possible to explicitly consider the contribution to heat transfer of large blood vessels either by specification of the vessel's wall temperature or by solution of the conjugated problem for the blood flow in the vessel. The latter could be accomplished using the Navier–Stokes equations of the viscous incompressible flow or assuming that analytic model of the Poiseuille flow is accurate enough for the problem in question. As a rule, specific features of the blood rheology are disregarded.

A simple description of the heat flux q from the blood with the temperature T_b flowing through the vessel of diameter d to the cylindrical region (diameter D) of tissue having the temperature T is possible via the overall heat transfer coefficient h as

$$q(s) = h\pi d(T(s) - T_b)\,,$$

where s is the coordinate along the vessel centerline. This coefficient is expressed via the so-called conduction shape factor $h\pi d = \lambda_t\sigma$, where

$$\sigma = \frac{2\pi}{\ln \frac{D}{d}}.$$

Similarly, the heat exchanged between the adjacent arteries and veins with the countercurrent flow is written with the corresponding conduction shape factor

$$\sigma_\Delta = \frac{2\pi}{\cosh^{-1}\left(\frac{w^2 - r_a^2 - r_y^2}{2r_a r_v}\right)}.$$

Similar relation is used for the description of heat transfer inside the blood vessels:

$$q(s) = h\pi d(T_w - T_b),$$

where T_w is the temperature of the vessel's wall.

The value of this coefficient is frequently determined from the empirical relations Nu = Nu(Gz) between the Nusselt number

$$Nu = \frac{hd}{\lambda_b}$$

and the Graetz number

$$Gz = \frac{\rho_b c_b u_0 d^2}{\lambda_b L_0}.$$

The relations Nu = Nu(Gz) are usually obtained by approximation of the experimental data for the flow of the Newtonian incompressible fluids in the straight pipes of the circular cross section. The blood vessels in living organisms are more complex objects, to say nothing about the blood rheology. Evidently, heat transfer description based on the shape conduction factors (determined also for other geometrical arrangements such as vessel near the skin surface or near a junction of vessels) is rather crude and more sophisticated simulation is necessary.

Large blood vessels embedded into the tumor region through the contribution to heat transfer by blood flow could lead to the significant temperature nonuniformity around the vessel and possible thermal underdosage. Moreover, heat transfer along the vessel may result in the damage of the surrounding healthy tissues. This problem is common for cryosurgery and hyperthermia with an obvious change between "cooling" and "heating." Crezee and Lagendijk and Kolios et al. were the first to study this issue. FD computations by Shih et al. for the cylindrical tissue sample with the embedded straight blood vessel coaxial to the sample showed that blood flow effect is significant when the vessel diameter is about 1 mm or larger.

Kolios et al. used FD method to compute the temperature distribution in the cylindrical blood vessels and in the surrounding tissue shaped as a coaxial cylinder of a larger diameter. The authors showed, inter alia, that the contribution of the blood flow in the central vessel to the global heat transfer in the considered region does depend on the blood perfusion level in the tissue (modeled as a source term in the Pennes equation (53)). Kolios et al. in the subsequent paper reported the results of the experimental study of the temperature gradients in the biological tissues (porcine kidney) near large vessels heated either locally via the needle perfused with the hot water or using a distributed ultrasound source. The authors have measured the temperature gradients up to 6 °C mm^{-1} and found that the effect of the vessel on the heat transfer primarily depends on its orientation with respect to the heat source. Some of the authors' conclusions are hardly unexpected - for example, the observation that the temperature gradients caused by the local heating are larger than those caused by the distributed heat source.

Kotte et al. suggested to consider discrete vessels (more exactly, segments of the vessels) as geometrical objects that could be characterized by a 3D parametric curve with associated diameters and consider a simplified analytical model of the heat transfer along the vessel, neglecting the logtitudinal diffusion of heat and assuming circular cross section. It is also suggested to characterize blood by the so-called *mixing cup temperature* that is defined as

$$T_{mix}(u)\pi R^2 = \int_0^R u(r)T(r)2\pi r\,dr,$$

where $T(r)$ is the real blood temperature, and define the heat exchange between the tissue and the blood using Nusselt number.

Experimental imitation of the heat transfer around the blood vessel embedded into tissue was performed by Zhang et al. who used massive copper block prefrozen in liquid nitrogen to freeze the gel volume that contained two tubes with the countercurrent warm water flow. The author have also derived an analytic solution for the heat transfer in the cylindrical tissue with a single embedded vessel.

Deng and Liu have considered two models, with a single artery transiting the tumor (SATT) and with counter-current flow vessels transiting the tumor (CVTT). The authors have neglected the conduction along the blood flow inside the vessel and assumed that heat transfer between the blood and surrounding tissues could be described by the heat transfer coefficient $h = \mathrm{Nu}k_b/D$, where D is the diameter of the vessel and the Nusselt number Nu is considered to be constant. Then the temperature of the blood T is governed by the equation:

$$\rho_b c_b \frac{\partial T}{\partial t} = \frac{hP}{S}(T_w - T) - \rho_b c_b \upsilon \frac{\partial T}{\partial z} + Q_r,$$

where T_w is the wall temperature of the vessel, $h = Nuk_b/D$ is the convective heat transfer coefficient, P and S are the perimeter of the vessel and its cross-sectional area, υ is the mean blood velocity along the vessel (in the direction z).

Mohammed and Verkey in their study of the effects of heat transfer by large blood vessels such as carotid artery in the neck region during LITT use incompressible Navier–Stokes for the flow of blood that was assumed to behave as Newtonian fluid:

$$\rho \frac{\partial u(r,t)}{\partial t} - \mu \nabla^2 u(r,t) + \rho u(r,t) \nabla u(r,t) + \nabla p = F, \quad \text{...(55)}$$

$$\nabla \cdot u(r,t) = 0 \quad \text{...(56)}$$

The authors additionally have assumed (not stating it explicitly) that the dynamic viscosity μ is constant.

The nonstationary form of the Navier–Stokes equations allows to simulate time-periodic blood flow, accounting for the beat cycle effect. For the vessels far away from the heart, the pumping cycle could be disregarded and the stationary Navier–Stokes equations used:

$$-\mu \nabla^2 u(r) + \rho u(r) \nabla u(r) + \nabla p = F, \quad \text{...(57)}$$

$$\nabla \cdot u(r) = 0. \quad \text{...(58)}$$

When the blood flow in the vessel is computed directly, it is usually assumed that the flow entering the region of interest is equilibrated and could be described by the well-known Poiseuille velocity profile, which for the 3D simulation is written as

$$u(x,y,0) = 2u\left(1 - \frac{x^2 + y^2}{R^2}\right).$$

There are examples of the application of the rigorous mathematical methods to the development of vascular models. Deuflhard and Hochmuth applied the homogenization approach to the multiscale periodic problem, which provide the description of the microvascular tissue. The authors assumed that the regions of blood are separated. Later this restriction was removed, allowing analysis of more realistic biologically problems.

Lubashevsky and Gafiychuk characterize living tissues as a hierarchically organized, active, heterogeneous medium that could be represented by a composition of two subsystems: (1) uniform cellular system and (2) highly branching hierarchical vascular network involving arterial and venous sub-networks.

The vascular network is embedded into the cellular tissue and in spite of its small relative volume determines to a large extent heat transfer in the organism due to the convective heat transport by the blood flow. Details of the vascular network could vary from patient to patient even for the same tissue.

In living tissues the arterial bed is typically located in the immediate vicinity of the corresponding venous bed. The blood temperature in arteries can differ significantly from the blood temperature in veins, thus there must be an essential heat exchange between an artery and the nearest vein of the same level. This effect is phenomenologically taken into account in the effective conductivity model.

Lubashevsky and Gafiychuk considered a hierarchical system of the interacting vessels of the different level (size); it was shown, inter alia, that the blood velocity could nonlocally depend on the temperature distribution in the tissue.

Temperature Fluctuations in Living Tissues

In all the models discussed in the preceding sections it was implicitly assumed that the blood perfusion is stationary. A model accounting for the pulsative character of the perfusion was formulated by Deng and Liu. The authors use the Pennes' equation (53), considering the case of the constant thermal diffusivity of the tissues and neglecting the external source:

$$\rho c \frac{\partial T}{\partial t} = \lambda \nabla^2 T + c_b \Omega (T_a - T) + \dot{q}_{\text{met}},$$

and represent all variables as a sum of a mean and a fluctuation:

$$T = \bar{T} + T', \quad T_a = \overline{Ta} + T_a', \quad \Omega = \bar{\Omega} + \Omega', \quad \dot{q}_{\text{met}} = \overline{\dot{q}_{\text{met}}} + \dot{q}_{\text{met}}',$$

where the mean values denoted by overline are averaged over the time interval Γ

$$\overline{A(\mathrm{r},t)} = \frac{1}{\Gamma}\int_{t-\Gamma/2}^{t-\Gamma/2} A(\mathrm{r},\tau)d\tau.$$

Evidently, the averaged values of fluctuation should be zero:

$$\overline{T'} = 0, \quad \overline{T_a'} = 0, \quad \overline{\Omega'} = 0, \quad \overline{\dot{q}_{\mathrm{met}}'} = 0.$$

The authors follow the well known fluid dynamics procedure (the derivation of the Reynolds equation for the turbulent flow) to write out the equation for the mean temperature:

$$\rho c\frac{\partial \overline{T}}{\partial t} = \lambda\nabla^2\overline{T} + c_b\overline{\Omega}(\overline{T_a} - \overline{T}) + c_b\overline{\Omega' T_a'} - c_b\overline{\Omega' T'} + \overline{\dot{q}_{\mathrm{met}}} \qquad ...(59)$$

and for the temperature fluctuations:

$$\rho c\frac{\partial T'}{\partial t} = \lambda\nabla^2 T' + \Omega' c_b(\overline{T_a} - \overline{T}) - c_b\overline{\Omega' T_a'} + c_b\overline{\Omega' T'}$$
$$-c_b\Omega' T' - c_b\overline{\Omega} T' + c_b\overline{\Omega} T_a' + c_b\Omega' T_a' + \dot{q}_{\mathrm{met}}'. \qquad ...(60)$$

Two additional terms in (59) $c_b\overline{\Omega' T_a'}$ and $c_b\overline{\Omega' T'}$ represent the mean transferred energy due to the perfusion perturbations and temperature fluctuations in the arterial blood and in the tissue, respectively. Assuming the mean quantities to be $O(1)$ while the fluctuations to be $O(\delta)$, it is possible to neglect in (59) all terms smaller than $O(1)$ and in (60) all terms smaller than $O(\delta)$ to get the final equations:

$$\rho c\frac{\partial \overline{T}}{\partial t} = \lambda\nabla^2\overline{T} + c_b\overline{\Omega}(\overline{T_a} - \overline{T}) + \overline{\dot{q}_{\mathrm{met}}} \qquad ...(61)$$

and for the temperature fluctuations

$$\rho c\frac{\partial T'}{\partial t} = \lambda\nabla^2 T' + \Omega' c_b(\overline{T_a} - \overline{T}) - c_b\overline{\Omega} T' + c_b\overline{\Omega} T_a' + \dot{q}_{\mathrm{met}}'. \qquad ...(62)$$

Exact Solutions to the Bioheat Equation

The exact solutions to the one-dimensional equation that governs heat transfer in tissues for the multi-block domain for the case of constant thermophysical properties were obtained by Durkee et al. in the Cartesian and spherical coordinates and by Durkee and Lee for the cylindrical coordinates; the nonstationary sources and nonstationary boundary conditions were considered by Durkee and Antich and the sinusoidal in time heat flux into the half-plane by Shih et al. An analytic solution of one-dimensional steady-state Pennes' bioheat transfer equation in cylindrical coordinates was also presented by Yue et al. The solution of the two-dimensional problem of the heat transfer of a pair of vessels in the finite medium were written out as a series expansion by Shrivastava and Roemer.

Solution of One-Dimensional Multiregion Bioheat Equation

Durkee et al. rewrite the classical Pennes equation (53) for the case of one-dimensional composite tissue having region-wise uniform thermomechanical properties. The equation governs the temperature evolution $T_m(r, t)$ in each region m that occupies the interval $R_{m-1} < r < R_m$ for $m = 1, 2, \ldots, M$ and expesses as

$$\rho_{tm} C_{pm} \frac{\partial T_m}{\partial t} = k_m \frac{1}{r^G} \frac{\partial}{\partial r}\left(r^G \frac{\partial T_m}{\partial r} \right) + m_{bm} C_{bm} (T_{am} - T_m) + Q_m + P_m, \quad \ldots(63)$$

where $G = 0, 1, 2$ correspond to the Cartesian, cylindrical, and spherical geometry, respectively.

The authors suggested to refer to the (63) as to the multiregion bioheat equation (MBE).

The equations in different regions are coupled by the continuity of the temperature

$$T_m(R_m, t) = T_{m+1}(R_m, t) \quad \ldots(64)$$

and of the heat flux

$$-k_m \left. \frac{\partial T_m}{\partial r} \right|_{r=R_m} = -k_m \left. \frac{\partial T_{m+1}}{\partial r} \right|_{r=R_m} \quad \ldots(65)$$

at the interfaces between the regions.

Different boundary conditions (Dirichlet, Neumann or other) could be specified at the external boundaries of the multiregion domain. For example, in the case of spherical symmetry the boundness of the temperature at the centre could be required:

$$\lim_{r \to 0} T_1(r, t) < \infty.$$

At the outer surface of the sample the authors used the so-called conductive–convective heat transfer condition:

$$-\hat{k}_M \left. \frac{\partial T_M}{\partial r} \right|_{r=R_m} = h(T_M(R_M, t) - T_\infty).$$

Here h and T_∞ are the convective heat transfer coefficient at the outer surface and the ambient temperature, respectively. $\hat{k}_M$ is not a true thermal conductivity but just a boundary condition coefficient.

The initial condition is a trivial one:

$$T_m(r,0) = T_m^0(r).$$

Durkee et al. introduce dimensionless variables for the temperature,

$$\psi_m(\rho,\tau),$$

position (and, hence, region boundaries),

$$\rho = r/R_M, \quad \rho_m = r_m/R_M,$$

and time,

$$\tau = \frac{k_D}{\rho_{tD} c_{pD} R_M^2} t,$$

to obtain the dimensionless form of MBE,

$$\frac{\partial \psi_m}{\partial \tau} = K_{1,m} \frac{1}{\rho^G} \frac{\partial}{\partial \rho} \left(\rho^G \frac{\partial \psi_m}{\partial \rho} \right) - K_{2,m} \psi_m + L_m,$$

and of the interface conditions for the continuity of the temperature and the heat fluxes

$$\psi_m(\rho_m,\tau) = \psi_{m+1}(\rho_m,\tau), \quad k_m \frac{\partial \psi_m}{\partial \rho}\bigg|_{\rho=\rho_m} = k_{m+1} \frac{\partial \psi_{m+1}}{\partial \rho}\bigg|_{\rho=\rho_m}.$$

The boundary conditions at the outer boundaries and the initial condition are transformed similarly.

Two family of constants $K_{1,m}$ and $K_{2,m}$ named by the authors *dispersion* and *removal* constants, respectively, are defined as

$$K_{1,m} = \frac{k_m \rho_{tD} c_{pD}}{k_D \rho_{tm} c_{pm}}, \quad K_{2,m} = \frac{m_{bm} c_{pbm} R_M^2 \rho_{tD}}{c_{pD}} k_D \rho_{tm} c_{pm}.$$

Heat source is then expressed as

$$L_m = \frac{\rho_{tD} c_{pD}}{\rho_{tm} c_{pm}} \frac{R_M^2}{k_D T_r^0} (m_{bm} c_{pbm} T_{arm} + Q_m + P_m).$$

The subscript "D" in the above expression indicate the reference region.

The authors additionally introduced the dimensional homogeneous H_1 and inhomogeneous H_2 heat transfer coefficients:

$$H_1 = h, \quad H_2 = -h \frac{T_\infty}{T_r^0}$$

and conduction coefficients $k_m = k_m/R_m$.

To solve the governing equations that are linear and separable, Durkee et al. decompose the solution into the transient ψ_m^{tr} and stationary parts ψ_m^s:

$$\psi_m(\rho,\tau) = \psi_m^{tr}(\rho,\tau) + \psi_m^s(\rho).$$

The former satisfies the following equation:

$$\frac{\psi_m^{tr}}{\partial \tau} = K_{1,m} \frac{1}{\rho^G} \frac{\partial}{\partial \rho}\left(\rho^G \frac{\psi_m^{tr}}{\partial \rho} \right) - K_{2,m}\psi_m^{tr}, \qquad \ldots(66)$$

while the steady state solution satisfies the equation

$$0 = K_{1,m} \frac{1}{\rho^G} \frac{\partial}{\partial \rho}\left(\rho^G \frac{\psi_m^{s}}{\partial \rho} \right) - K_{2,m}\psi_m^{s} + L_m.$$

The solution is sought as a series expansion

$$\psi_m^{tr}(\rho,\tau) = \sum_{l=1}^{N_\alpha} R_{m,i}(\rho)\exp(-\alpha_l^2 \tau)$$

where the eigenfunctions are determined from the equation

$$\frac{1}{\rho^G} \frac{1}{d\rho}\left(\rho^G \frac{dR_{m,i}}{d\rho} \right) + \left(\frac{\alpha_i^2 - K_{2,m}}{K_{1,m}} \right) R_{m,i} = 0.$$

The eigenvalues α_i are region independent (hence the subscript "m" is omitted).

Details of the derivation could be found in the cited papers. Durkee et al. presented the eigenfunctions for the cases of the Cartesian and spherical geometry that are expressed via sine and cosine and hyperbolic sine and cosine. In the cylindrical geometry the solution is written out using the Bessel and modified Bessel functions.

Numerical evaluation of the obtained solution was performed by Durkee et al. in the subsequent paper for the one-, two-, four-, five-, and eight- region problems. The authors reported no comparison with the experimental data and stresses that they consider their solutions as the benchmarks only for the verification of the numerical approaches to the solution of the bioheat equations.

Transient heat sources and transient boundary conditions have been considered by the first two of these authors. Two forms of the heating term $P_m(t)$ and of the ambient temperature T_∞ were analyzed. The first one is the stepwise source and temperature variation that can be described as

$$P_m(t) = \sum_{l=1}^{L} H_P(t - t_{m,l})\Delta P_{m,l}$$

for the sources localized in the specific region and

$$T_\infty(t) = \sum_{k=1}^{K} H_T(t - t_k)\Delta T_{\infty,k}$$

for the ambient temperature. Here H_P $(t - tm,\ l)$ and H_T $(t - t_k)$ are Heaviside functions. The second form considered is the sinusoidal perturbation of the ambient temperature

$$T_\infty(t) = T_\infty^{av} + T_\infty^{a}\sin(\Omega t),$$

where T_∞^{av} is the average ambient temperature, T_∞^{a} is the amplitude of its perturbation, and Ω is the frequency.

The authors proceed similar to the previously considered case of stationary source and boundary conditions with some modifications. The separation of the solution into the nonsteady ψ_m^{tr} and steady ψ_m^{s} parts is written as

$$\psi_m(\rho,\tau) = \psi_m^{tr}(\rho,\tau) + \psi_m^{s}(\rho)H_2(\tau)$$

where in contrast to the steady state case is an explicit time dependence due to the variation of the ambient temperature introduced via the nonhomogeneous surface heat transfer coefficient H_2 (τ) that is defined as

$$H_2(\tau) = -h\frac{T_\infty(\tau)}{T_r^0}.$$

For the transient part, one then have instead of (66) the following relation:

$$\frac{\psi_m^{tr}}{\partial\tau} = K_{1,m}\frac{1}{\rho^G}\frac{\partial}{\partial\rho}\left(\rho^G\frac{\psi_m^{tr}}{\partial\rho}\right) - K_{2,m}\psi_m^{tr} + \hat{S}_m(\tau),$$

where the source term is written as

$$\hat{S}_m(\tau) = S_m(\tau) - \left[\psi_m^{s}\frac{dH_2(\tau)}{d\tau}\right].$$

Performing rather lengthy calculations, the authors wrote out the solution in terms of the multi-regional Green function $G_{m,n}$:

$$\psi_m^{tr} = \sum_{n=1}^{M}\left[\int_{\rho_{n-1}}^{\rho_n}\rho'^G G_{m,n}(\rho,\tau,\rho',\tau')\Big|_{\tau'=0}\psi_m^{tr}(\rho,0)\,d\rho'\right.$$

$$\left.+\int_0^{\tau}\int_{\rho_{n-1}}^{\rho_n}\rho'^G G_{m,n}(\rho,\tau,\rho',\tau')\hat{S}_n(\tau')d\rho' d\tau'\right].$$

The expression for the Green function $G_{m,n}$ (ρ,τ,ρ',τ') itself could be found in the cited paper. As in the steady state case, the authors note that the intended use of their results is as the benchmarks for the numerical methods verification only and does not try to compare computations with experimental data or draw general conclusions on the peculiarities of the heat transfer in tissues relevant to the physiologic problems.

Heat Transfer with a Sinusoidal Heat Flux on Skin Surface

The problem of the temperature response of the semi-infinite sample of the biological tissue to periodic (sinusoidal) heat flux applied to its surface was considered by Shih et al. This problem was earlier considered by Liu and Xu who were interested, however, not in the transient response, but in the long run steady periodic temperature oscillations and in the phase shift between the heat flux and the surface temperature that could be utilized for diagnostics. The mapping of the temperature of the skin surface as a diagnostic tool was also considered by Deng and Liu.

One-dimensional Pennes equation (53) is considered:

$$\rho_t c_t \frac{\partial T}{\partial t} + w_b c_b (T - T_a) = k \frac{\partial^2 T}{\partial x^2}.$$

The oscillatory heat flux with the amplitude q_0 and frequency Ω is given at the boundary $x = 0$

$$-k \frac{\partial T}{\partial x} = q_0 e^{i\Omega t}.$$

Introducing the dimensionless variables:

$$z \equiv \Omega t,\ \alpha \equiv \frac{k}{\rho_t c_t},\ X \equiv \frac{\sqrt{\Omega}}{\alpha},\ c_1 \equiv \frac{w_b c_b}{\rho_t c_t \Omega},\ \theta \equiv \frac{k(T - T_a)}{q_0} \frac{\sqrt{\Omega}}{\alpha}.$$

Shih et al. get the dimensionless form of the equation

$$\frac{\partial \theta}{\partial z} + c_1 \theta = \frac{\partial^2 \theta}{\partial X^2}$$

and of the boundary condition

$$\frac{\partial \theta}{\partial X} = -e^{iz}.$$

Using the Laplace transform method, the authors finally obtained solution for the temperature evolution as

$$\theta = \frac{e^{iz}}{2\sqrt{c_1 + i}} \left[e^{-\sqrt{c_1 + i}X} \operatorname{erfc}\left(\frac{X}{2\sqrt{Z}} - \sqrt{(c_1 + i)z} \right) \right.$$

$$\left. - e^{\sqrt{c_1 + i}X} \operatorname{erfc}\left(\frac{X}{2\sqrt{z}} + \sqrt{(c_1 + i)z} \right) \right]$$

and for the temperature response on the skin surface as

$$\theta(X=0)=\frac{e^{iz}}{\sqrt{c_1+i}}\operatorname{erf}\left[\sqrt{(c_1+i)z}\right].$$

Using the obtained solutions, the authors studied the effect of the frequency of the heating force on the affected depth of the tissues and discussed implications for the optimal choice of the boundary heat flux for the measurements of blood perfusion rate.

Freezing of the Cylindrical Tissue Region with a Single Embedded Coaxial Blood Vessel

This problem considered by Zhang et al. is stated as follows. The long (so the end effects could be neglected) cylindrical sample of the tissue contains a single cylindrical blood vessel that is coaxial to the sample. The steady state case is studied with the thermal properties of the tissue being constant throughout the frozen and unfrozen tissue subregions. The exterior cylindrical boundary of the tissue ($r = b$) is maintained at the constant freezing temperature T_0, and the boundaries $z = 0$ and $z = L_t$ are adiabatic. The authors suppose that this problem could serve a crude model of the blood vessel deep inside the frozen tumor region. Similar problem relevant to the hyperthermia under a few additional assumptions was considered by Kotte et al.

The equation for the dimensionless tissue temperature $T^* = (T_t - T_0)/(T_a - T_0)$, where T_a is the fixed body core temperature, is written as

$$\frac{1}{r}\frac{\partial}{\partial r}\left(r\frac{\partial T^*(r,z)}{\partial r}\right)+\frac{\partial^2 T^*(r,z)}{\partial z^2}=0$$

and the boundary conditions as

$$T^*(r,z)=0 \quad \text{at} \quad r=b,$$
$$T^*(r,z)=f(z) \quad \text{at} \quad r=a,$$
$$\frac{\partial T^*(r,z)}{\partial z}=0 \quad \text{at} \quad z=0, L_t,$$

where $f(z)$ is the temperature along the vessel wall. Under assumption that the blood rheology could be approximated by the viscous Newtonian fluid, the Poiseuille velocity profile

$$u=u_0\left(1-\left(\frac{r}{a}\right)^2\right)$$

was used for the flow in the vessel, where u_0 is the average blood velocity over the vessel cross-section. The equation for the dimensionless blood temperature T_b^* is formulated as

$$\frac{\lambda_b}{\rho_b C_b}\frac{1}{r}\left(\frac{\partial}{\partial r}\left(r\frac{\partial T_b^*}{\partial r}\right)\right) = u_0\left(1-\left(\frac{r}{a}\right)^2\right)\frac{dT_c}{dz},$$

where λ_b is the blood thermal conductivity, ρ_b is the blood density, C_b is the blood specific heat, and T_c is the blood bulk temperature defined as

$$T_c = \frac{4}{a^2}\int_0^a T_b\left(1-\left(\frac{r}{a}\right)^2\right)rdr.$$

The thermal resistance of the wall of the blood vessel is neglected; at the boundary between the blood and tissue regions the continuity of both the temperature and the heat flux is required.

Using the separation of variables, the authors sought the tissue temperature distribution in the form

$$T^*(r, z) = R_0(r)\, Z(z),$$

with $Z(z)$ satisfying

$$\frac{d^2 Z(z)}{dz^2} + \beta_m^2 Z(z) = 0$$

and $R_0(r)$ satisfying

$$\frac{d^2 R_0(r)}{dr^2} + \frac{1}{r}\frac{dR_0(r)}{dr} - \beta_m^2 R_0(r) = 0.$$

Here β_m are the positive roots of the equation $\sin(\beta L_t) = 0$, that is, $\beta_m = m\pi/L_t$.

The final solution is expressed as

$$T^*(r,z) = A_0 \ln\left(\frac{r}{b}\right)\cos(\beta_0 z) + \sum_{m=1}^{\infty} A_m\left[-\frac{K_0(\beta_m b)}{I_0(\beta_m b)}I_0(\beta_m r) + K_0(\beta_m r)\right]\cos(\beta_m z).$$

The coefficients A_m are determined as follows:

$$A_0 = \frac{1}{\ln\left(\frac{a}{b}\right)L_t}\int_0^{L_t}\cos(\beta_0 z)f(z)dz,$$

for $m = 0$ and

$$A_m = \frac{2\int_0^{L_t}\cos(\beta_m z)f(z)dz}{\left[\frac{K_0(\beta_m b)}{I_0(\beta_m b)}I_0(\beta_m a) + K_0(\beta_m a)\right]L_t},$$

for $m = 1,\ldots$

Heat Transfer in a Finite Tissue Region with two Embedded Blood Vessels

Shrivastava and Roemer considered analytically heat transfer in the uniform tissue region with two embedded blood vessels and found the so-called Poisson conduction shape factors for both the vessel–vessel heat transfer (VVPCSF) and the vessel–tissue heat transfer (VTPCSF) that could be used for relatively low cost computations of heat transfer in the tissue. Some of the energy leaving the blood vessel goes to the other vessel; the rest goes to the tissue boundary that can be expressed as

$$Q_{total', v1} = Q_{v1\to v2} + Q_{v1\to t'},$$
$$Q_{total, v2} = Q_{v2\to v1} + Q_{v2\to t'}.$$

Mathematically the problem reduces to the solution of 2D Poisson equation in the domain that contains circular holes, if conduction is the only heat transfer mechanism.

Such problem has been considered earlier by Zhu and Weinbaum for the case of a single hole embedded into a finite medium and by Wu et al. for two holes in the infinite medium. The vessels of diameter ≈100– 400 μm make the majority of the counter-current vessels pairs in organism that significantly influence heat transfer in tissues.

Shrivastava and Roemer considered the circular domain with two circular holes of smaller (not necessarily equal) diameter. The heat source P is assumed to be uniform in the tissue part of the domain. Starting with the Poisson equation for the temperature distribution

$$\frac{1}{R}\frac{\partial}{\partial R}\left(R\frac{\partial T}{\partial R}\right)+\frac{1}{R^2}\frac{\partial^2 T}{\partial \psi^2}=P,$$

with Dirichlet conditions specified on all boundaries, the authors by substituting

$$T = T_1 - \frac{PR^2}{4}$$

convert it into the Laplace equation:

$$\frac{1}{R}\frac{\partial}{\partial R}\left(R\frac{\partial T_1}{\partial R}\right)+\frac{1}{R^2}\frac{\partial^2 T_1}{\partial \psi^2}=0.$$

where P is the power deposition.

Using the linearity of the governing equations, the authors subdivide the original problem of the disk with two eccentric holes into two subproblems for the disk with a single hole. These latter problems are further simplified by the conformal transformation that makes the disk and the hole to be co-centric. The final equation for the first subproblem reads as

$$\frac{1}{R_1^*}\frac{\partial}{\partial R_1^*}\left(R_1^*\frac{\partial T_{11}}{\partial R_1^*}\right)+\frac{1}{R_1^{*2}}\frac{\partial^2 T_{11}}{\partial \alpha_1^2}=0,$$

where R_1^* is the dimensionless radius in the transformed coordinate system and α_1 is the corresponding angular coordinate. The general solution to this equation is expressed as

$$T_{11}=A_{01}+A'_{01}\ln(R_1^*)+\sum_{n=1}^{\infty}[A_{n1}(R_1^*)^n+A'_{n1}(R_1^*)^{-n}]\sin(n\alpha_1)$$
$$+\sum_{n=1}^{\infty}[B_{n1}(R_1^*)^n+B'_{n1}(R_1^*)^{-n}]\cos(n\alpha_1).$$

Similar calculations are performed to obtain the analytic solution to the second subproblem for T_{12}.

Once the temperature field is determined, the heat transfer rates Q are found by integration, for example, $Q_{v1\to t}$ is expressed as

$$Q_{v1\to t}=-\int_0^{2\pi}\left.\frac{\partial T}{\partial R}\right|_{R=R_{v1}}R_{v1}d\theta,$$

where θ is the angular position relative to the coordinate system with origin at the vessel's center.

The analytical solution procedure developed by the authors could be extended to the general case of N vessels embedded into the tissue with their arbitrary location by decomposition into N subproblems for the single embedded vessel.

Heat Transfer in Cryosurgery

With a few exceptions, analysis of the cryosurgery problems is performed using numerical models, which, as was already mentioned, should be verified and validated. The most popular working fluid (phantom material) in experiments designed for the verification of the numerical approach is gelatin solution that inhibits convection in the fluid domain and thus imitate the situation in real biological tissues.

The finite difference method for the computation of the two-dimensional temperature field around the cryoprobe was used for the first time probably by Cooper and Trezek. Deng and Liu described the boundary element method for the computation of the temperature field in two- and three-probes cryosurgery systems. Two-dimensional computations were performed using the finite element method and the finite volume method on unstructured triangular grids. An axisymmetric problem of the ice ball formation around a single cryoprobe was solved in three-dimensional formulation in the Cartesian coordinates using the finite element method. The anisotropic adaptation of the computational grid suggested by the authors in the earlier papers was used. Linearized equations were solved by the preconditioned (by the incomplete factorization method ILU) GMRES method. Software components from the PETSC library were used.

The accuracy of the heat transfer computations increase if it is possible to monitor the evolution of the crystallization front; in this case the boundary elements method are applicable. Restoration of the domain of the cryoaction from the experimental data is a separate problem. Usually one gets a number of 2D parallel image slices that should be processed to obtain 3D shape of the tumor. The reconstruction of 3D surface from a sequence of 2D images is the noncorrect problem and additional information should be used such as smoothness of the transition between adjacent input shapes and smoothness of the reconstructed 3D surface; different measures of the smoothess of the involved geometric entities could be chosen.

Solution of this ill-conditioned problem in cryosurgery is greatly simplified if a priori information on the shape and evolution of the frozen domain is available. Unfortunately, a considerable part of this information, for example, the assumption that the crystallization front is a simply connected convex single boundary surrounding the probe, is valid for the single-probe cryosurgery systems only.

Heat Transfer in the Single Probe Cryosurgery System

The main distinction of finite difference computations performed by Cooper and Trezek from the traditional engineering heat transfer analysis was the use of the bioheat equation for the unfrozen tissue region. These authors computed the stationary temperature field around the cryoprobe tip that had either spherical or the hemispherical shape in the axisymmetric formulation. They also considered one-dimensional approximation for the semispherical cryoprobe that predicts for the steady state radius of the ice ball the following expression:

$$r^* = \frac{\sqrt{\beta}-1}{2\sqrt{\beta}} + \left[\frac{(1-\sqrt{\beta})^2}{4\beta} + \frac{1+\phi}{\sqrt{\beta}}\right]^{1/2},$$

where $r^* = R/r_0$ is the dimensionless ice ball radius related to the probe radius r_0, λ_0 and λ_1 are the thermal conductivities of the unfrozen and frozen tissues, respectively, T_0 is the deep body temperature, T_1 is the temperature of the cryoprobe tip, and

$$\beta = \frac{\rho_b c_b r_0^2}{\lambda_0}, \quad \phi = -\frac{\lambda_1}{\lambda_0}\left[\frac{T_m - T_1}{T_m - T_0}\right].$$

A bit later, Comini and Del Giudice implemented in the finite element context the enthalpy methods for the solution of the phase change problems by the introduction of the enthalpy variable $H = \int_{T_0}^{T} \rho c \, dT$, which includes the latent heat effect. They have considered a hemispherical cryoprobe and a flat disc-shaped probe.

Rabin and Shitzer extended the finite difference method developed earlier by Rabin and Korin for the nonbiological engineering problems to bioheat equation and applied it to the computation of the temperature distribution around the spherical and cylindrical cryoprobes. The essential improvement consisted in the implicit approximation of the perfusion term in (53) that increase stability of the scheme in the regions with the high perfusion rates and allowed to increase the time integration step. The radius of the spherical probes considered were 5, 15, and 25 mm (the latter is hardly relevant to the practical cryosurgery systems) and from 2 to 7 mm for the cylindrical probes. The authors found that the cooling power by the cryoprobe is highly nonuniform in time, being greatest at the earlier stage and then tending to the constant value. They also estimated the mass flow rate of nitrogen needed to provide the required cooling power.

The temperature distribution around a single cryoprobe was computed, among other problems, by Fortin and Belhamadia. The method used is called by the authors semi-phase field formulation, since it is intermediate between the enthalpy method and the phase field approach. To be more precise, instead of the differential equation for the phase field variable two algebraic equations are solved. The necessary spatial resolution is achieved by the anisotropic mesh adaptation, key element for which is the edge refinement with the edge length specification using a solution dependent metric.

Heat Transfer in the Multiprobe Cryosurgery System

The success of the cryosurgery operation depends on the correct prescription of the cooling and thawing rates and on the exact localization of the domain of freezing that provide both the liquidation of malignant tissues and the preservation of healthy organs intact. It is not easy to accomplish due to the following reasons:

1. As a rule, the cooling/thawing protocol is determined by the general considerations and rarely takes into account the patient-specific anatomic data (the size and the geometrical shape of the tumor,

the degree of the development of the capillary system, the closeness of the walls of important organs, etc.).

2. It is difficult to optimally place a large number of cryoprobes (and, probably, a cryoheater) in space using the surgeon's experience alone.
3. Methods of the surgery monitoring (the ultrasound diagnostics, the thermometric, the matrix thermovision, the X-ray computer tomography, the NMR tomography, the optical coherent tomography, the electric impedance tomography) do not provide the sufficiently complete information on the three-dimensional temperature field; the temperature distribution inside the ice ball could be determined either by the extrapolation based on the mathematical models or by the NMR tomography using the empirical correlation between the temperature and the measurements results in Hounsfield units (HU) normalized by the contrast water-air; the accuracy of this approach is limited due to the sensitivity to the noise and the nonmonotonous character of the correlation dependence. The typical resolution of, for example, ultrasound imaging is not better than 1 mm. It should be also noted that the temperature of the leading front of the ice ball may be well below 0 °C (e.g., –8 °C during cryoablation of the prostate).

It is relatively easy to determine the position of the boundary of the ice ball (e.g., using MR imaging due to the excellent contrast between ice and surrounding tissues), but not of the necrosis domain that usually has a volume equal to one-fourth of the volume of the ice ball. The required temperature should be attained with a safe margin around the tumor, which is about 1 cm for the hepatic tumor and about 2–3 mm for the prostate cancer. The Cryoablation Simulator (CryoSim) developed initially to support cryosurgery of the prostate (but could be used for other organs as well) relies on the transformed ultrasound images of the ice ball and the simplified model of heat transfer for the case of the constant thermal properties of tissues.

The value of the simulation is determined by the accuracy of its predictions. Rabin systemized the sources of errors in the cryosurgery problems; one of the most important is the insufficient knowledge of the thermophysical properties of the living tissues at low temperatures.

Zhao et al. used finite element analysis implemented in ANSYS commercial software to study processes in a novel minimally invasive probe system capable of performing both cryosurgery and hyperthermia. The tissues were treated as nonideal materials freezing over a temperature range, the thermophysical properties of which are temperature dependent. The enthalpy method is applied to solve the heat transfer problem with the phase change.

The heat transfer in the tissues when 3-, 6-, or 18-probe cryosystems were simulated. One of the aims of this study was the analysis of the sensitivity of the temperature field to the character of the temperature dependence of the effective thermal conductivity of the frozen tissues (this problem was earlier treated analytically for the one-dimensional case). The stationary problem was considered that corresponds to the exposition stage of the cryosurgery operation. In this case the temperature distribution and the position of the phase boundary depend on the thermal conductivity only and do not depend on the specific heat of the tissues and on the latent heat of the phase transition. Note that the stationary two-dimensional formulation is the *worst case scenario* in respect to the preservation of the surrounding healthy tissues and organs.

To derive the discrete equations, the finite volume method on unstructured triangular grids was used. The linearized equations were solved by the stabilized bi-conjugated gradient method Bi-CGSTAB.

There is a number of approaches to the generation of the unstructured grids: the recursive division, the advancing front method and its modification for the generation of the nonuniform grids with node density increase towards the surface – the advancing layer method, and the Delauney triangulation. The latter was used as a basic one in the cited papers The grid generation procedure includes three steps:

1. Triangulation of the boundary notes by the Green–Sibson algorithm
2. Addition of new nodes that improve the grid quality locally
3. Grid optimization by local swapping of the edges of the neighbor elements, smoothing of the grid bonds using the spring analogy, and selective removal of *bad* nodes. The void formed after the node deletion is triangulated by the variant of the advancing front method.

Control of the cryoaction domain in the multiprobe systems

The use of multiple cryoprobes allows more flexible control of the extension of the freezing domain.

The unstructured computational grid (left) and the fragment of the temperature field (right) for the three-probe cryosurgery system (probes are positioned in the vertices of the equilateral triangle with the side equal to 27 mm).

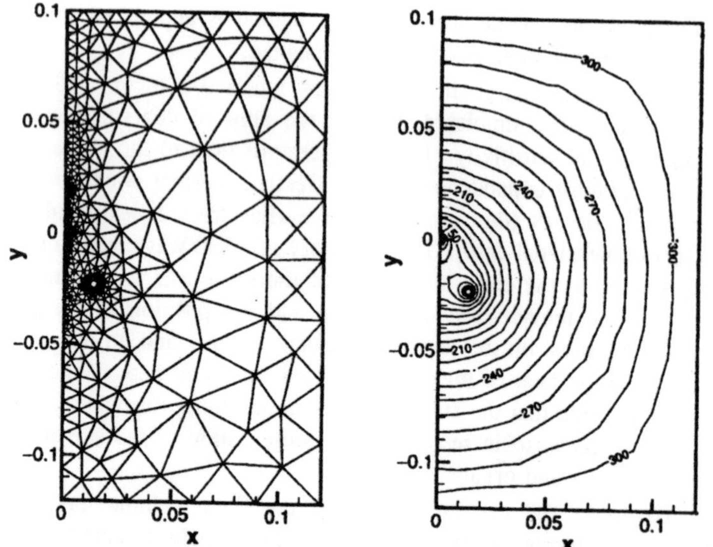

Fig. 8.4. Computational grid (left) and the temperature field (right) for the three-probe cryosurgery system.

The computational grid for the whole domain (left) and the temperature field near the cryoprobes (right) for the 18-probe system. The cryoprobes are placed in the three concentric layers. The central boundary with temperature 37 °C could be considered either as a cryoheater surface or as the large blood vessel.

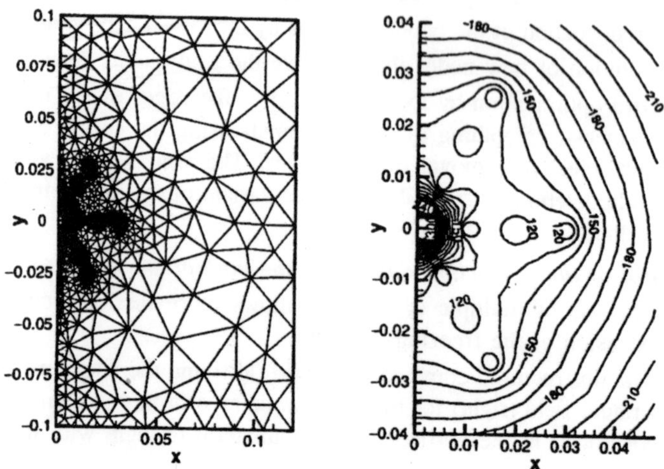

Fig. 8.5. Computational grid (left) and the temperature field (right) for the 18-probe cryosurgery system.

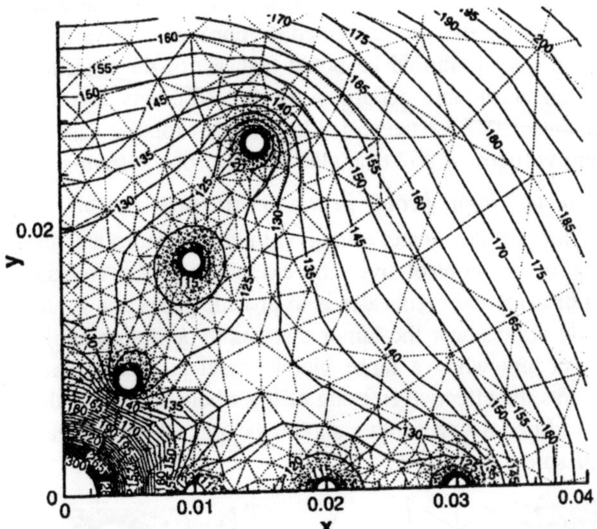

Fig. 8.6. Computational grid and the temperature field for the 18-probe cryosurgery system.

It could be seen that the freezing zones around the individual cryoprobes merge. The one-dimensional horizontal and vertical distributions of the temperature allows one to judge the enhancement of the control over the freezing domain with increase of the number of cryoprobes in the system (solid and dashed lines correspond to the systems having three and 18 probes).

The thermal load of the cryoprobes in the 18-probe system differs greatly. The cryoprobes of the inner shell provide the greatest heat flux from the tissues and, thus, require the greater flow rate of the cooling fluid (either liquid nitrogen or argon, depending on the cryosurgery system design). The cryoprobes in the middle shell, as

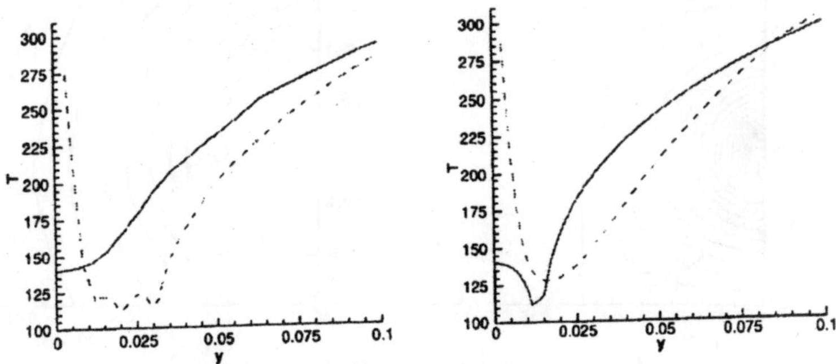

Fig. 8.7. One-dimensional temperature distributions for 3- and 18-probe systems.

could be easily seen from the isotherms, cool the tissue with very low efficiency. Evidently, the presented cryoprobes disposition is far from the optimal one. It should be noted that use of the unstructured grids provides an acceptable accuracy for the relatively coarse grids: the grid contains 1,044 vertices, 1,876 triangles for the case of 3 probes and 4,869 vertices, 8,967 triangles for 18-probe system.

Use of an additional heater in the cryosurgery systems

The use of the heater along with the cryoprobes increases the flexibility of the control over the extension of the freezing domain. The aim of using the cryoheater is to provide protection to the important tissues and organs near the tumor that has to be destructed. The similar situation occurs when a large blood vessel is found near the tumor that prevents the effective freezing of the whole domain of the malignant tissue. Additional cryoprobes should be placed in this case near the vessel to shield its action.

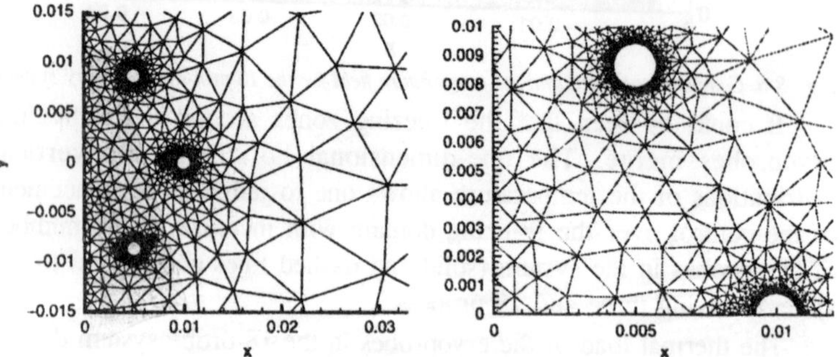

Fig. 8.8. Computational grid for 6-probe system.

The computational grid and the temperature field for the six-probe cryosurgery system without the cryoheater are presented in Figs. 8.8

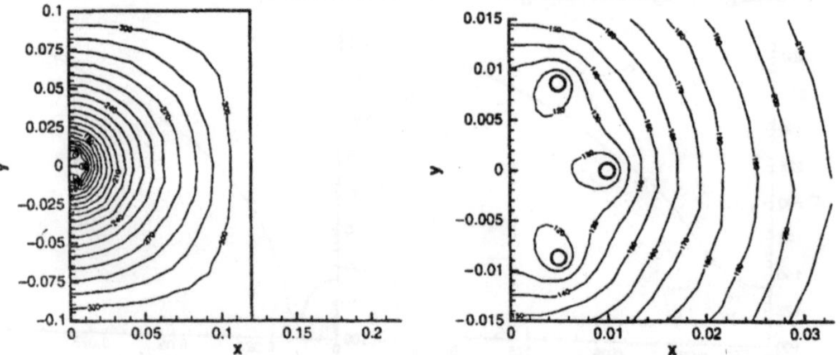

Fig. 8.9. Temperature field for 6-probe system.

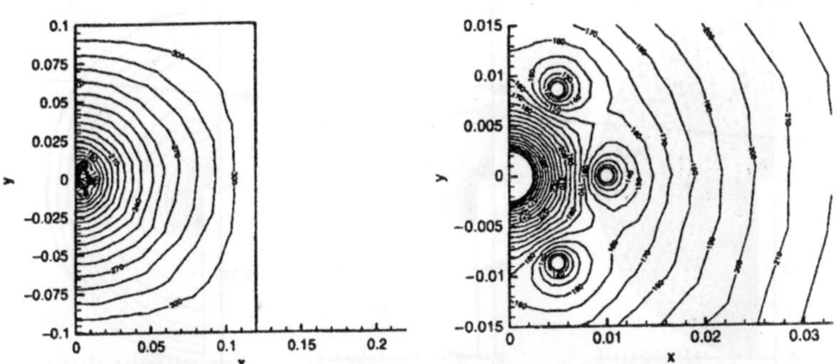

Fig. 8.10. Temperature field for 6-probe system with cryoheater.

and 8.9, respectively. The corresponding temperature field for the same system with an additional cryoheater is shown in Fig. 8.10 and the comparison of the temperature profiles in Fig. 8.11.

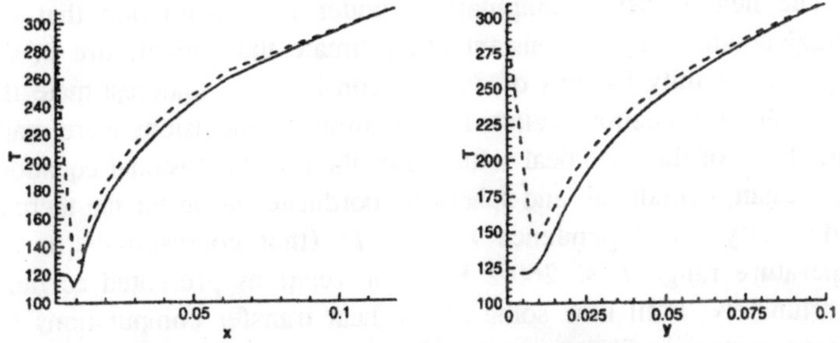

Fig. 8.11. Temperature distribution for 6-probe system without and with cryoheater.

It could be seen that indeed the use of the cryoheater with the cryosurgery system enhances the flexibility of the control and improves the localization of the freezing domain. It should be stressed, however, that the price for the flexibility in the freezing control are the greater temperature gradients (and, hence, the greater level of the thermoelastic stresses).

Effect of the tissue thermal conductivity

An example of the heat transfer computations for a single cryoprobe (an axisymmetric problem) for the cases of the constant (center) and temperature dependent (right) thermal conductivity is presented in Fig. 8.12. The following dependence for the effective tissue thermal conductivity ($\mathrm{Wm^{-1}\,K^{-1}}$) has been used:

$$k(T) = \begin{cases} K(T), & T < T_1 \\ K(T_1) + (k_2 - K(T_1))(T - T_1)/(T_2 - T_1), & T_1 \le T \le T_2 \\ k_2, & T > T_2 \end{cases}$$

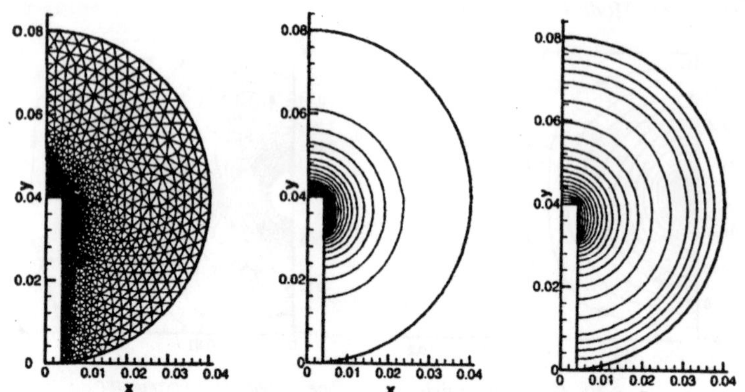

Fig. 8.12. Temperature field around a single cryoprobe.

$T_1=260.2K$, $T_2=273.2K$, $k_2=0.49$, $K(T)=k_0/T^{\alpha}$, $k_0=2,135$, $\alpha=1.235$.

The heat transfer computations under the assumption that the thermal conductivity is constant overestimate the temperature in the cryoprobe vicinity by tens of °C and considerably underestimate the size of the cryoablation region. Earlier similar conclusions were made on the basis of the analytical solutions of the one-dimensional equations in Cartesian, cylindrical, and spherical coordinates using for the thermal conductivity the dependence $k = k_0/T^n$ (that corresponds to the temperature range $T < 260.2$ K in the relations presented earlier). Unfortunately, until now some of the heat transfer computations for the cryosurgery problems are performed for the constant or piece-wise constant thermal conductivity of the tissues.

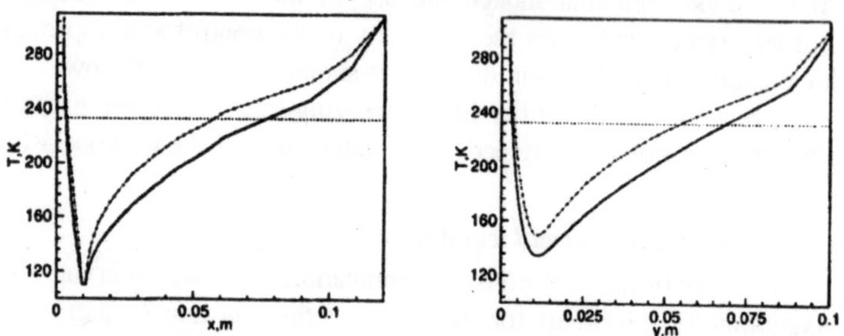

Fig. 8.13. The temperature distribution along the lines x=0 and y=0 in the 6-probe system: solid, the variable thermal conductivity and dashed, the constant thermal conductivity.

In the computations presented (as well as in other studies as far as the author knows), the thermal conductivity of the frozen tissue depends only on temperature and usually is assumed to be equal to

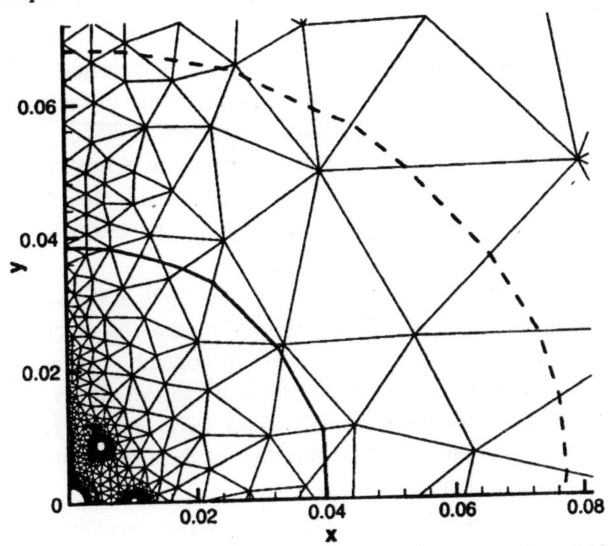

Fig. 8.14. Comparison of the ice ball size for the constant and variable thermal conductivity: 6-probe cryosystem.

that of the ice. The accuracy of this assumption is hard to estimate a priori. The two cases are considered: (1) the constant thermal conductivity (note that its value does not present in the formulation of

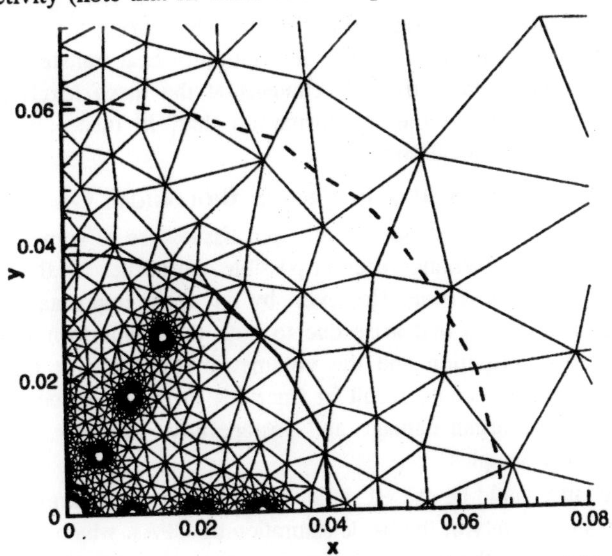

Fig. 8.15. Comparison of the ice ball size for the constant and variable thermal conductivity: 18-probe cryosystem.

the stationary problem) and (2) the temperature-dependent thermal conductivity obtained by the matching of conductivity of the water and the ice. Computations for the cryosurgery systems containing up to 18 cryoprobes showed that the indeterminacy of the temperature field could be as high as several dozens of degrees centigrade and a few centimeters in the extent of the cryoablation zone estimated by the value of the critical (*lethal*) isotherm.

Development of the adequate model of the thermal conductivity of the biological tissues as the heterogeneous media that includes ice, extracellular matrix, water - vitrified or bonded - is a separate problem. An approach based on the traditional methods for the effective properties of the composite materials is not sufficient since both the fraction of the unfrozen water and the state of the extracellular matrix vary during the cryoaction.

It is well known that the first cycle - freezing–exposition–thawing - results in the increase of the thermal conductivity of the tissues and in the considerable (dozens of per cents) increase of the volume of the ice ball for the second cycle. The destruction of the cellular membranes and the blood vessels are considered as the possible mechanisms of these alterations.

An adequate modeling of the heat transfer during the cryoaction thus should combine macroscopic description of heat transfer with the analysis of the media structure changes at the meso level such as cells dehydration and the ice formation that, in turn, are mainly determined by the local cooling.

Mechanical Stress in Frozen Biological Objects

The mechanical stresses in unfrozen tissues are of some interest, for example, for assessing different possible approaches for the process of the cryoprobe insertion. However, by far, more important stress is in the frozen tissues that arise due to both the nonuniformity of the temperature distribution and the volumetric expansion at the phase change, since they could result in fractures in frozen tissues causing the irreversible organ damage and postoperative bleeding, or in the cryopreserved samples.

The latter could be caused, in addition to the temperature gradients, by the nonuniformity of the tissue saturation by CPAs, which are known to have different thermal expansion coefficients. The danger of the crack appearance in the vitrified samples is also related to the risk of devitrification, since defects could serve as nucleation sites.

Stress in Frozen Tissues

In all studies of stresses in cryobiology known to the author, it is assumed that the phase interface between the frozen and unfrozen tissues is sharp. Certainly, it is a good approximation in many cases, but still an approximation. In reality, the mushy zone of finite thickness exists that provide a smooth transition from the liquid to the solid phase. This zone, its mechanical properties, and relation to the crystal state were studied in the metallurgical context using two phase models - solid porous medium saturated with liquid - by Mo and collegues.

Rabin and Steif in a number of publications considered the problem of the stress distribution in the spherical ice ball around a cryoprobe. The problem is one-dimensional and analytical solution could be obtained. The solution derived by the authors does not arise doubts, but the problem formulation does. There are other researches that also assume that the just transformed from the liquid to the solid material should have the same stress state as liquid; in other words, the stress at the advancing crystallization front should be purely dilatational, that is, deviatoric stress should be zero. The origin of this misunderstanding is unknown to the author. Evidently, it contradicts the atomistic theory of the structure of liquids and solids. The atom, molecule, or other growth unit (e.g., in protein crystallization) has to be incorporated into the existing crystal lattice. There is no physical law that forbid the new crystalline layers to be in the deformed state. It is well known, for example, that growth of the strained layers is common in the semiconductor processing: this occurs in growth in semiconductor heterostructures with different lattice constants from the vapor or the liquid phase. If the reader is not satisfied with these arguments, there is another proofreductio ad absurdumprovided by the supporters of the *zero deviatoric stress* notion themselves. The boundary condition for the circumferential (tangential) stress component on the surface of the spherical ice ball requires this component to be zero during growth, while allows an arbitrary value during melting with the instantaneous switch when the growth stops and the crystal begin to melt. This means that at that moment the stress distribution is changed (stress field is governed by the elliptic equations) and the elastic energy of crystal is instantly changed - and that is impossible in our Universe.

He and Bishop solved the axisymmetric problem and found the stress distribution in kidneys during the cryosurgical operation. These authors used a correct boundary condition: the surface of the ice ball was free to move.

Dennis et al. in their optimization of the cryopreservation procedure solved 3D elasticity problem, which for the constant material properties in Cartesian coordinates is written as

$$(\lambda+G)\left[\frac{\partial^2 u}{\partial x^2}+\frac{\partial^2 \upsilon}{\partial x\partial y}+\frac{\partial^2 w}{\partial x\partial z}\right]+G\nabla^2 u+X=0,$$

$$(\lambda+G)\left[\frac{\partial^2 u}{\partial x\partial y}+\frac{\partial^2 \upsilon}{\partial y^2}+\frac{\partial^2 w}{\partial y\partial z}\right]+G\nabla^2 \upsilon+Y=0,$$

$$(\lambda+G)\left[\frac{\partial^2 u}{\partial x\partial z}+\frac{\partial^2 \upsilon}{\partial y\partial z}+\frac{\partial^2 w}{\partial z^2}\right]+G\nabla^2 w+Z=0,$$

where

$$\lambda=\frac{E\nu}{(1+\nu)(1-2\nu)},\quad G=\frac{E}{2(1+\nu)}.$$

The body forces per unit volume due to the thermal expansion/contraction over the temperature interval D are

$$X=-(3\lambda+2G)\frac{\partial(\beta T)}{\partial x},Y=-(3\lambda+2G)\frac{\partial(\beta T)}{\partial y},Z=-(3\lambda+2G)\frac{\partial(\beta T)}{\partial z}.$$

Stress in Vitrified Biological Objects

Stress development and relaxation in the glass forming liquids during crystal growth could affect both the nucleation and growth. Stress development is attributed to the differences in the density of the liquid and solid phases. Stress relaxation is determined by the nature of the glassy state and usually described by the some relaxation law. For example, the so-called Maxwellian relation is governed by the following equation:

$$\varepsilon=\varepsilon_0\exp\left(-\frac{\tau_G}{\tau_R}\right),$$

where ε is a parameter that defines (by multiplication with the number of particle incorporated into the crystal) the change of the energy of the elastic deformation caused by addition of the new particles to the growing crystal, τ_G is the time required for the formation of one crystalline layer, and τ_R is the characteristic relaxation time; the former is determined by the crystal growth rate as

$$\frac{d_0}{\tau_G}=\frac{D}{4d_0}\left[1-\exp\left(-\frac{\Delta\mu-\varepsilon}{kT}\right)\right].$$

Here d_0 is the characteristic size of the particle, D is the diffusion coefficient, and $\Delta\mu$ is the difference (per particle) of the chemical potentials in the liquid and solid phases.

The authors have found that below the so-called decoupling temperature (named so because viscous relaxation and molecular motion decouple), the stresses could have a considerable effect on the growth process.

Steif et al. studied stress in the vitrified CPA solutions in the absence of the tissue samples to study the vitrification ability of CPA themselves, which are thought to be dominant also in case of the tissue vitrification as well as in partially vitrified biological objects.

Experiments showed that the patterns of fractures in CPA solutions depend on the solution composition and the cooling rate; fractures disappear on thawing. The authors note that it is difficult to distinguish between the recrystallization and devitrification in the experimental study and proposed a new term *rewarming phase crystallization* to unite these phenomena.

It was found that fractures occur in the temperature interval from −150 to −120 °C during warming. Both CPA contained in the vial and CPA droplet were studied. The droplet could deform elastically, viscously, and due to the thermal contraction. Calculation of the radial and circumferential stress distributions performed in the cited paper were based on the Vogel–Fulcher–Tammann temperature dependence for the solution viscosity under the assumption that temperature is uniform. The behavior of the CPA solution was modeled with a viscoelastic law of the form

$$\dot{\varepsilon} = \frac{1}{E}\frac{d\sigma}{dt} + \frac{\sigma}{3\eta} + \beta\frac{dT}{dt}, \qquad \text{...(67)}$$

where $\dot{\varepsilon}$ is the strain rate, σ is the stress, E is the Young modulus, β is the coefficient of the thermal expansion. This model represents the strain rate as a sum of three contributions: the elastic strain rate, the viscous strain rate, and the thermal strain rate.

The notion of the *set temperature* introduced for the description of transition from the viscous to the elastic strain regime. This temperature is defined as the temperature at which effects of the viscous strain rate and elastic strain rate are comparable:

$$P \equiv 3\frac{d\eta}{dT}\frac{B}{E} \approx 1,$$

where B is the cooling rate.

For the simple symmetric shapes the temperature distribution can be found analytically from (46). The maximal temperature difference is

$$\Delta T = f\frac{Bd^2}{\alpha},$$

where d is the size of the sample, α is the thermal diffusivity, and coefficient f is equal to 1/8 for the plate, 1/16 for the cylinder, and 1/24 for the sphere.

The maximal tensile stress

$$\sigma_{max} = g\frac{E\beta\Delta T}{1-\nu}$$

corresponds to the in-plane for plate, the axial stress for the cylinder, and the circumferential stress for the sphere and in all cases is located at the geometrical center. Here ν is the Poisson ration and the coefficient g is equal to 1/3, 1/2, and 2/5 for the plate, cylinder, and sphere, respectively [114].

If the specimen is likely to fracture when the current stress σ exceeds the yielding stress $\sigma_{crit,}$ the critical size could be defined for the given cooling conditions:

$$d_{crit} = \sqrt{\frac{1-\nu}{fg}\frac{\alpha}{\beta B}\frac{\sigma_{crit}}{E}}.$$

Partial vitrification

Steif et al. considered also a more complex situation of partial vitrification. They deal with the case when CPA solution vitrifies while crystallization occurs in the preserved sample. Partial vitrification could also occur in the biological tissue during conventional cryopreservation when the residual solution becomes so concentrated that could vitrify in the presence of ice.

Authors analyzed two cases:

1. Straight long blood vessel in one of three configurations:
   (a) The inner wall of the vessel is covered by the thin film of the CPA solution
   (b) The inner space of the vessel is filled with the CPA solution
   (c) The whole of the vessel is immersed in the larger cylindrical container filled with the CPA solution
2. Curved blood vessel placed in the storage container and completely covered with CPA solution; the centerline of the vessel forms a 270° arc.

Axisymmetric problem was solved for the straight vessel and 3D one for the curved vessel. CPA is modeled as a viscoelastic fluid in accordance with (67) and the blood vessel itself as an elastic medium below the freezing temperature.

Steif et al. found that stress in the CPA region decreases with increase of the solution volume. Planar cracks are predicted to form in the radial direction and propagate to the blood vessel inner wall. Potentially such cracks could damage the endothelial cells.

Optimization of Cryoaction

Suboptimal cryosurgery may leave cells in the tumor nontreated, may lead to the damage of the healthy surrounding tissues, may require unnecessary large number of cryoprobes, and increase the duration of the operation as well as increase the risk of the post surgery complications. To formulate an optimization problem one should specify the following:

1. Optimization (control) parameters p = $(p_1, p_2, \dots, p_n)$
2. A goal function F(p).

The choice of control parameters in cryosurgery is evident: (1) a temperature (to be more exact, its variation in time – the cooling and thawing rates and the duration of the exposition ("holding")); (2) number of cryoprobes; (3) their positions. Sometimes (and always in cryopreservation) additional parameters define saturation of tissues by cryoprotectant and its removal (usually a tuple of three values – the concentration, the temperature, and the time interval) for all the process stages).

A stepped cooling with a long exposition at the intermediate temperature is often used in cryoconservation to provide the sufficient time for the extracellular ice formation and dehydration of cells with a subsequent fast cooling. Sometimes this procedure is referred to as Interrupted Slow Freezing (ISF). If during the slow cooling step a critical concentration of CPA in the cell is reached, the intracellular solution will be vitrified during the fast freezing. To determine the intracellular water volume and the solution concentration in the cell at different temperatures, Liu et al. used the following equations:

$$\frac{dV_w}{dT} = -\frac{ARTL_p}{\upsilon_{10} B_1}\left[\ln\left(\frac{V_w}{V_w + (n_n^i + n_s^i)\upsilon_{10}}\right) + \ln\left(1 + \frac{G'C(T)}{100 - C(T)}\right)\right], \quad ...(68)$$

$$\frac{dn_s^i}{dT} = \frac{AP_{CPA}(T)}{B_1}\left(\frac{1000 R_t C}{MW_s(100 - C)(R_t + 1)} - \frac{n_s^i}{V_w}\right), \quad ...(69)$$

where A is the area of the cell surface, B_1 is the slow cooling rate, C is the extracellular concentration of CPA, n^i_s is the number of moles of solute in the cell, L_p is the hydraulic permeability, υ_{10} is the molar volume of water, P_{CPA} is the solute permeability, R is the universal gas constant, MW is the molecular weight, the subscripts "w," "s," and "n" refer to the water, solute, and salt, respectively, and

$$G' = \frac{R_t}{1+R_t} MW_w \left(\frac{1}{MW_n R_t} + \frac{1}{MW_s} \right),$$

R_t is ratio of CPA (DMSO) to NaCl by weight. It is assumed that both the hydraulic permeability and the CPA permeability vary with temperature by the Arrhenius-type dependence.

The authors used these equations to find optimal (i.e., providing the critical level of the CPA in the cell while avoiding intracellular ice formation) cooling protocols and used mouse embryos to confirm the predictions experimentally.

Note that traditional specification of the cooling regimes by the piecewise constant cooling rate is not optimal due to the considerable nonlinearity of the tissues' response.

Cryoprotectant unloading in several steps is shown to increase the post- thaw fertilization rate of oocytes, probably due to avoiding the rupture of the fragile plasma membranes by the shearing pressure of the water rushing out of the cell to the hypotonic environment in the case of the single step unloading.

It is more difficult to write the goal function. In the radiation therapy a radiation dose or a probability of destruction of the pathological tissue (based on the approximation of experimental data) are used and corresponding goal functions are called physical and biological. Since reaction of tissue depends on the dose that surrounding tissues get, a histogram *dose-volume* is also exploited.

Planning of the hyperthermia treatment is usually based on the quadratic function defining the deviation of the computed temperature field T from the desired (optimal) one T_0:

$$F(p) = \int_{\Omega} (T - T_0)^2 dV.$$

Sometimes instead of the averaged values, one or several control points in the tumor region are considered.

It is not necessary to restrict the tumor temperature from above. One could just specify T_d – the minimal temperature that provides the destruction of pathological tissues. Additionally one should specify a "safe" temperature value – the maximal acceptable temperature for the healthy tissues T_{lim}.

In the solution of the optimization problem

$$\min F(p), \quad F(p) = \int_{\Omega_{\text{tumor}}, T<T_d} (T - T_d)^2 dV,$$

restrictions on the temperature $T \leq T_{\text{lim}}$ in the region of the healthy tissue $\Omega \backslash \Omega_{\text{tumor}}$ could be either used explicitly as box type constraints or introduced into the goal function using the penalty approach:

$$F(p) = \int_{\Omega_{\text{tumor}}, T<T_d} (T - T_d)^2 dV + w \int_{\Omega/\Omega_{\text{tumor}}, T \geq T_{\text{lim}}} (T - T_{\text{lim}})^2 dV.$$

where w is the weight factor.

Similar approaches are used in cryosurgery. The simplest one to specify a goal function is to define a critical (*lethal*) temperature that provide the tissue destruction

$$F(p) = \int_{\Omega} w(T) dV,$$

where the weight function $w(T)$ equals to 0 or 1 in the corresponding temperature intervals and the geometrical regions. However, this *lethal* temperature does depend on the tissue type and other factors, including the cooling/thawing protocol; the reported value differs greatly from –2 °C for the dog osteocytes down to –70 °C for rat adenocarcinoma. Berger and Poledna, for example, assumed the lethal temperature to be equal to –50 °C in their optimization analysis of the optimal probe placement in cryosurgery of the liver.

Baissalov et al. optimized the cryoaction for the six-probe system. To cope with the inertia and nonlinear response of the cryoprobes, calibration measurements were performed and used to parameterize the temperature of the cryoprobe's tips. During the "stick" period (a clinical procedure to get the cryoprobe adhered to the tissue) $0 < t < t^{st}$, it was assumed to be equal to –10 °C. During the freezing period (the cryosurgical operation itself), the temperature history was approximated by the two matched exponential functions:

$$t^{st} < t < t^* : T_i(t) = T_i^P (1 - b_{1i} \exp[-a_1 (t - t^{st})]),$$

where $b_{1i} = (T_i^P - T_i^{st}) / T_i^P$ and

$$t^{st} < t < t^{off} : T_i(t) = T_i^P (1 - b_{2i} \exp[-a_2 (t - t^{st})]),$$

$b_{2i} = (T_i^P - T_i^*) / T_i^P$.

The parameters, including the temperature of dependence matching t^*, were found by fit the experimental thermal histories. The objective (goal) function was formulated as

$$F_T = \sum_{i=1}^{n}\left[w\left(T^{pr} - T_i\right)^2 + \alpha S_T\left(T_i - U^{pr}\right)\right] + \sum_{j=1}^{l}\sum_{i=1}^{m_j}\beta_j S_{N,j}\left(L_{ij}^{pr} - T_i\right),$$

where T^{pr} is the prescribed temperature in anatomical region of interest (ROI), l is the number of the normal tissue region near the tumor, n and m are the number of the points where the temperature is prescribed for the tumor and jth normal tissue, respectively, T_i is the actual temperature, α is the penalty factor for exceeding the given upper temperature limit in the tumor U^{pr}, β is the penalty factor for exceeding the given lower temperature limit in the healthy tissue, w is the weighting factor, and S_T and $S_{N,j}$ are the step functions given as

$$S_T = \begin{cases} 1, & U^{pr} - T_i < 0 \\ 0, & U^{pr} - T_i > 0 \end{cases}$$

and

$$S_{N,j} = \begin{cases} 1, & L^{pr} - T_i > 0 \\ 0, & L^{pr} - T_i < 0 \end{cases}$$

These authors in the subsequent paper had named the goal function presented above as *quadratic temperature* form and also considered *linear volume* form:

$$F_V = \sum_{i=1}^{I_T} S_i^T (P_i^T - V_i^T) + \sum_{j=1}^{l}\sum_{i=1}^{I_N} S_{i,j}^N (V_i^{N,j} - P_i^{N,j}),$$

where I_T and I_N are the number of isotherms considered for the tumor and the normal tissues, respectively, l is the number of the normal tissue regions, P and V with the corresponding sub- and superscripts are the prescribed and computed volumes of the tissues completely enclosed by the ith isotherm.

Weighted combination of the goal functions

$$F_C = w_P F_P + w_T F_T$$

also was considered.

In total, 24 variables (polar angle, radius, target temperature, and freeze time for each cryoprobe) were simultaneously optimized using the Broyden– Fletcher–Goldfarb–Shanno algorithm with an approximate reconstruction of the Hessian in the course of iterations; the needed gradients of the goal function were determined numerically. The linear volume form was found the fastest to converge; two other forms of the goal functions, however, provide better solution.

Index of freezing exposition is introduced as a product of the exposition time at temperature lower than a prescribed one by the average temperature during this time interval:

$$I = \int_{T_1}^{T_2} T(t)dt.$$

Unfortunately, these criteria are not universal ones. For example, one of the optimization goals could be minimization of the operation duration.

In cryopreservation optimization, the goal function could be based on the specified level of the thermoelastic stresses that arise in the preserved sample. Dennis et al. optimized the cooling rate of the cryopreservation procedure under the following conditions:

1. The cooling rate should be minimal.
2. The local stresses in the tissue should not exceed the specified level σ_{yield}.

The goal function was formulated as

$$F(p) = -\left[\frac{\Delta T}{\Delta t} + P\left(\frac{\sigma_{\max}}{\sigma_{\text{yield}}}\right)^2\right],$$

where σ_{max} is the maximal value of the stress in the ROI and P is the weight of the penalty term. Stresses were computed solving 3D elasticity problem in the Cartesian coordinates under assumption of the constant tissue properties.

Recently, Yu and Liu suggested to use the entropy generation – the measure of the irreversibility of the process – as an index quantifying the cryosurgery injury.

The change of the entropy of the system dS is the sum of two contributions: the flow of entropy due to the interactions with the exterior dS_e and the entropy change due to the changes inside the system d_iS ($d_iS \geq 0$ according to the second principle of thermodynamics): $dS = d_eS + d_iS$ and the change of d_iS in time is expressed as

$$\frac{d_iS}{dt} = \int_V \sigma dV.$$

Here σ is the entropy generation per unit volume per unit time that is caused by the following sources:

1. Heat conduction
2. Mass diffusion
3. Gradients of the velocity field
4. Chemical reactions

and could be written as

$$\sigma = -\frac{1}{T^2} q \cdot \nabla T - \frac{1}{T} \sum_{i=1}^{n} J_i \cdot \left(T \nabla \frac{\mu_i}{T} - F_i \right) - \frac{1}{T} \Pi : \nabla v - \frac{1}{T} \sum_{j=1}^{r} J_j \cdot A_j .$$

Here q is the heat flux, J_i is the diffusion flux of the *i*th component, μ_i is the chemical potential of the *i*th component, F_i is the external force, Π is the viscous stress tensor, v is the velocity of the medium, and A_j is the chemical affinity of the reaction *j*.

Some of these processes are relevant for the cryosurgery, for example, the mass diffusion at the cellular level; however, the main contribution to the entropy generation is certainly heat transfer.

Yu and Liu proposed to compute the local entropy generation at the given position r for the specified time interval from τ_1 to τ_2

$$\Gamma_r = \int_{\tau_1}^{\tau_2} \sigma(r,\tau) d\tau$$

and for the whole region

$$\Gamma = \int_V \int_{\tau_1}^{\tau_2} \sigma(r,\tau) d\tau dV .$$

These quantities Γ_r and Γ as measures of the freezing injury could be used to define the goal function for the optimization of the cryoaction (maximize the entropy production in the tumor) as well as for specifying constraints (minimize the entropy production in the regions occupied by the healthy tissues).

9

PRESERVATION OF FOOD

Special attention must be paid to the possibility of contamination of food with toxic compounds. They may be present incidentally and may be derived in various ways. Examples of such contaminants are:

1. Pollutants derived from burning of fossil fuels, radionuclides from fallout, or emissions from industrial processing (toxic trace elements, radionuclides, polycyclic aromatic hydrocarbons, dioxins).
2. Components of packaging material and of other frequently used products (monomers, polymer stabilizers, plasticizers, polychlorinated biphenyls, cleansing/washing agents and disinfectants).
3. Toxic metabolites of microorganisms (enterotoxins, mycotoxins).
4. Residues of plant-protective agents (PPA).
5. Residues from livestock and poultry husbandry (veterinary medicinals and feed additives).

Toxic food contaminants might also be formed within the food itself or within the human digestive tract by reactions of some food ingredients and additives (e.g. nitrosamines). Measures required to prevent contamination include:

1. Extensive analytical control of food.
2. Determination of the sources of contamination.
3. Legislation (legal standards to permit, ban, curtail or control the use of potent food contaminants, and the processes associated with them) to establish permissible levels of contaminants.

Toxicological assessment of a contaminant may, for various reasons, be a difficult task. Firstly, sufficient data are not available for all compounds. Also, the possibility of synergistic effects of various substances, often including their degradation products, should not be excluded. Further, the hazard might be influenced by age, sex, state of health and by habitual consumption. Based on these considerations, any nutritional statement about the "*tolerable concentration*" must take sufficient safety factors into account.

Toxicity assay involves the determination of:

1. Acute toxicity, designated as LD_{50} (the dose that will kill 50% of the animals in a test series).
2. Subacute toxicity, determined by animal feeding tests lasting four weeks.
3. Chronic toxicity, assessed by animal feeding tests lasting 6 months to 2 years.

In chronic toxicity tests attention is especially given to the occurrence of carcinogenic, mutagenic and teratogenic symptoms. The tests are conducted with at least two animal species, one of which is not a rodent.

The upper dosage level for a substance, fed to test animals over their life span and observed for several generations, which does not produce any effect, is designated as the "No Observed Adverse Effect Level" (NOAEL, mg/kg body weight of the animal tested per day or mg/kg feed per day). This level can be used as a basis for estimating the hazard for humans in all cases in which a correlation between dose and effect has been observed. The NOAEL is multiplied by a safety factor (SF: 10^{-1} to 5×10^{-4}, mostly 10^{-2}), with which especially sensitive persons, extreme deviations from the average consumption and other unknown factors are taken into account, giving the toxicologically acceptable dose. It is expressed as the acute RfD (Reference Dose in mg/kg body weight (BW)/day) or as the Acceptable Daily Intake (ADI).

RfD. The acute RfD is the estimated amount of a chemical compound present in food which, related to the body weight and based on all the facts known at the time of the estimation, can be taken in over a short period of time (usually during a meal or a day) without posing a discernible risk to the consumer's health.

ADI. The ADI value denotes the amount of substance which the consumer can take in every day and lifelong with food without discernible injury to his health.

Taking the consumption habits into account, the RfD can be used to calculate the tolerable concentrations (TC) of substances for individual foods:

$$TC = \frac{NOAEL \times FV}{SF} \times \frac{BW}{CA \times ASF}$$

In this formula, TC is the toxicological tolerable concentration for a particular food (expressed in mg/kg food); NOAEL, no observed adverse effect level (mg/kg feed); FV, daily intake of feed by test animals (kg feed/kg body weight); SF, safety factor (10–2000, but usually 100); BW, body weight of an adult (50–80 kg); CA, amount in kg consumed per day of the food for which the TC is being calculated; and ASF, additional safety factor (up to 10) for particularly sensitive persons, such as children or the sick. The maximum concentrations of contaminants (MRL, maximum residue limit in mg/kg food) allowed by legislation are often still well below toxicological tolerance concentrations because other parameters such as "good agricultural practice" are taken into account.

The ADI value is compared with the NEDI or IEDI value (national or international estimated daily intake) to check the risk which comes from contaminants, e. g., pesticides. If the last mentioned value is higher than the ADI, tests are conducted to find out if there is in fact a risk, which would then lead to further measures, if necessary. In many countries, food monitoring is carried out for the early detection of possible danger due to undesirable substances like plant-protective agents (PPA), heavy metals and other contaminants. Foods belonging to the most important groups of goods are repeatedly tested for the presence of certain contaminants.

Toxic Trace Elements

Arsenic

From the viewpoint of its frequency in the environment, toxic activity and the probability of man's exposure to the substance, arsenic was first on the list of dangerous substances compiled in the USA in 1999. Arsenic was followed by lead, mercury, vinyl chloride, benzene, PCBs, cadmium and benzo[a]pyrene. The amount of arsenic which is probably not dangerous when taken orally is estimated at 0.3 μg/kg body weight/day.

Mercury

Mercury poisoning caused by food intake is derived from organomercury compounds, e. g., dimethyl mercury (CH_3—Hg—CH_3), methyl mercury salts (CH_3—Hg; X = chloride or phosphate), and phenyl mercury salts (C_6H_5—Hg X; X = chloride or acetate). These highly toxic compounds are lipid soluble, readily absorbed and accumulate in erythrocytes and the central nervous system. Some are used as fungicides and for treating seeds (seed dressing). Methyl mercury compounds are also synthesized by microflora from inorganic mercury salt sediments found on lake and river bottoms. Hence, the content of these compounds might rise in fish and other organisms living in water.

The natural mercury level in the environment appears to have stabilized in the last 50 years. Poisonings recorded in Japan appear to have been caused by consumption of fish caught in waters heavily contaminated by mercury-containing industrial waste water, and in Iraq by milling and consuming seed cereals dressed with mercury, which were intended for sowing. The tolerable dose for an adult of 70 kg is 0.35mg Hg per week, of which a maximum of 0.2 mg may be derived from the highly toxic methyl mercury.

Lead

The contamination of the environment with lead is increased by industrialization and by emissions from cars running on leaded gasoline. Tetraethyl-lead [$(C_2H_5)_4Pb$], an antiknocking additive used to increase the octane value of gasoline, is converted by combustion into PbO, $PbCl_2$ and other inorganic lead compounds. The major part of these compounds is found in an approx. 30 m wide band along roads or highways; the lead level sharply decreases beyond this distance. At a distance of 100 m from a road with heavy traffic, the lead level in the atmosphere decreases by a factor of 10 and that in soil and plants by a factor of 20 from the level found at or close to the road. A decrease in the level of lead in gasoline and increased use of unleaded gasoline has resulted in a drop in the extent of contamination. Environmental lead contamination has not, however, significantly increased the level of lead in food. The lead in soil is rather immobilized; thus the increase in the lead level of plants is not proportional to the extent of soil contamination. Vegetables with larger surface areas (spinach, cabbage) may contain higher levels of lead when cultivated near the lead emission source. When contaminated plants are fed to animals, the body does not absorb much lead since most is excreted in feces.

Further sources of contamination are lead-containing tin cookware and soldered metal cans and lead-containing enamels. This is particularly so in contact with sour food. These sources of contamination are of lesser importance. 1.75 mg of lead are considered as the tolerable weekly dose for adults of 70 kg.

Hair and bone analyses have revealed that lead contamination of humans in preindustrialized times was apparently higher than today. This might be due to the use in those days of lead pipes for drinking water, lead-containing tinware, and excessive use of lead salts for heavily glazed pottery used as kitchenware.

Cadmium

Cadmium ions, unlike Pb^{2+} and Hg^{2+}, are readily absorbed by plants and distributed uniformly in all their tissues, thus de-contamination by dehulling or by removal of outer leaves, as with lead, is not possible. Certain wild mushrooms (horse mushrooms, giant mushrooms etc.), peanuts and linseed can contain larger amounts of cadmium. The finer the linseed is ground, the higher the intake of cadmium on consumption. The contamination sources are industrial waste water and the sludge from plant clarifiers, which is often used as fertilizer.

A prolonged intake of cadmium results in its accumulation in the human organism, primarily in liver and kidney. A level of 0.2–0.3 mg Cd/g kidney cortex causes damage of the tubuli. The tolerable weekly dose for an adult (70 kg) is considered to be 0.49mg of cadmium. On the whole, the concentrations of the toxic trace elements lead, mercury and cadmium in food show a clearly decreasing tendency, especially in recent studies. This is partly due to improvements in trace analyses, but also due to a real decrease in food.

Radionuclides

It is estimated that the average radiation exposure in FR Germany in 1975 was 172 mrad, of which 21 mrad were ascribed to internal radiation by natural radionuclides incorporated in the body (about 90% from ^{40}K, the rest from ^{14}C) and less than 1 mrad by nuclides acquired as a result of atmospheric fallout from nuclear explosion tests (50% from ^{137}Cs, a radionuclide with a half life of 30 years, but quickly excreted by the body; approx. 50% from ^{90}Sr, a most dangerous radioisotope, capable of inducing leukemia and bone cancer; and traces of ^{14}C and tritium). ^{137}Cs and ^{90}Sr are the escort elements of potassium and calcium, respectively.

Food contamination with radionuclides in FR Germany had its peak in 1964/65, when the intake in food per day per person was 240 pCi of ^{137}Cs and 30 pCi of ^{90}Sr. Up to the Chernobyl reactor accident in April 1986, the intake was less than 10% of previous values as a result of the moratorium on atmospheric testing of atomic weapons. Radionuclide residues in food were not a health hazard.

For 1986, the accident in Chernobyl caused an additional intake of radionuclides with food that is estimated at (children up to the age of 1 year/adults) 1779/4598 Bq/year of ^{131}J, 986/1758 Bq/year of ^{134}Cs, and 1849/3399 Bq/year of ^{137}Cs. The resulting additional effective equivalent dose for people in the FR Germany is estimated at 0.06–0.22 mSv. In comparison, natural radiation exposure is about 2 mSv per year, of which 0.38 mSv/year is caused by radionuclides in food. As a precaution, maximum activity values of 500 Bq/l and 250 Bq/kg have been stipulated for milk and vegetables respectively. In comparison, the activity of natural radionuclides (mainly ^{40}K) in food is: milk 40–50Bq/kg, milk powder 400–500 Bq/kg, fruit juice concentrate 600–800 Bq/kg and soluble coffee (powder) >1000 Bq/kg.

The level of tritium infiltrating the biosphere is expected to rise further due to increasing nuclear plant operation worldwide.

Toxic Compounds of Microbial Origin

Food Poisoning by Bacterial Toxins

Most (60–90%) cases of food poisoning are bacterial in nature. They are distinguished by food intake causing:

1. Intoxication (poisoning, e. g., by *Clostridium botulinum*, *Staphylococcus aureus*).
2. Diseases caused by massive pollution with facultative pathogenic spores, e. g., *Clostridium perfringens*, *Bacillus cereus*.
3. Infections by *Salmonella* spp. or *Shigella* spp., *Escherichia coli*.
4. Diseases of unclear etiology, such as those from *Proteus* spp., *Pseudomonas* spp.

The harmful activity of these bacteria in the digestive tract is ascribed to enterotoxins, which are classified into two groups: exotoxins (toxins excreted by microorganisms into the surrounding medium) and endotoxins (retained by the microorganism cells but released when the cell disintegrates).

Exotoxins are released primarily by gram-positive bacteria during their growth. They consist mostly of proteins which are antigenic and very poisonous. They become active after a latent period. This group includes the toxins released by *Clostridium botulinum* (botulin toxin, a globular protein neurotoxin), *Cl. perfringens* and *Staphylococcus aureus*. Intoxications with *St. aureus* are the most frequent cause of food poisoning. Symptoms are vomiting, diarrhea and stomach ache and are caused primarily by food of animal origin (meat and meat products, poultry, cheese, potato salad, pastry).

Endotoxins are produced primarily by gramnegative bacteria. They act as antigens, are firmly bound to the bacterial cell wall and are complex in nature. They have protein, polysaccharide and lipid components. Endotoxins are relatively heat stable and are in general active without a latent period. The toxins causing typhoid and paratyphoid fevers, salmonellosis and bacterial dysentery are in this group. Salmonellosis is very serious. It is an infection by toxins of about 300 different but closely related organisms. The infection is characterized by enteric fever, gastroenteritis and salmonella septicemia. Sources of infections are egg products, frozen poultry, ground or minced beef, confectionery products and cocoa.

Although the bacterium *E. coli* first served only as an indicator of fecal contamination, it has meanwhile received special attention. This bacterium also includes enterotoxic strains, e. g., the especially dangerous strain 0157:H7 discovered in 1983.

Mycotoxins

There are more than 200 mycotoxins produced under certain conditions by about 120 fungi or molds. Infections of rye and, to a lesser extent, of other cereal grains with *Claviceps purpurea* (ergot, or rooster's spur) are responsible for the disease called ergotism (symptoms: gangrene and convulsions). The disease was important in

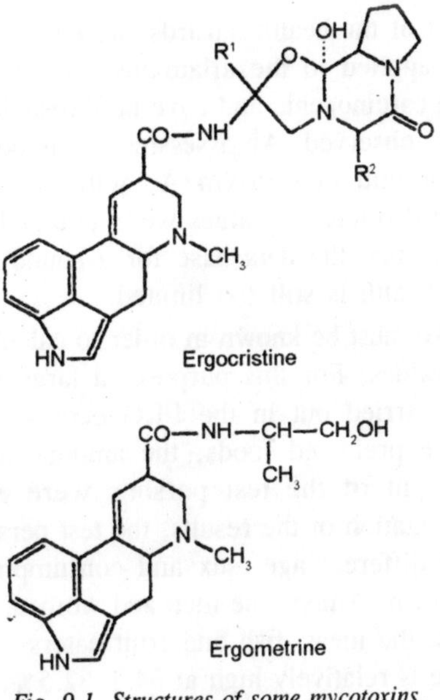

Fig. 9.1. Structures of some mycotoxins.

the past when bread from infected rye grain was eaten. It has practically ceased to exist due to seed treatment with fungistatic agents and grain cleaning prior to milling.

Most mycotoxin data are on the genera *Aspergillus* spp. and the aflatoxins they produce during growth. These are the most common and highly toxic fungal toxins, e.g., aflatoxin B_1, the most powerful carcinogen known. In animal feeding tests with rats, its carcinogenic effect is revealed at a daily dose of only 10 μg/kg body weight. In a comparative study, the carcinogenic property of the highly toxic dimethylnitrosamine was revealed only at a daily dose of 750 μg/kg body weight. It is primarily plant material (particularly nuts and´fruit) that is contaminated with aflatoxins. Aflatoxin passes from moldy feed to animal products, primarily milk. The dairy cow's metabolism converts the B-group aflatoxins to those of the M-group ("M" stands for metabolite), which are also carcinogenic. Nephrotoxic ochratoxin A passes from fodder cereals mainly to the blood and kidney tissue of pigs, but it is also found in the muscles, liver and adipose tissue.

In the course of food monitoring between 1995 and 2002, more than 40 foods were tested for the presence of aflatoxins, deoxynivalenol, fumosins, patulin, ochratoxin A und zearalenone. Individual mycotoxins were detected in 21% of the samples; pistachios were especially conspicuous.

An assessment of the health hazards caused by mycotoxins is not meaningful when applied to the aflatoxins because these substances damage DNA, are carcinogenic and have no threshold below which no harmful effects are observed. An assessment was possible in the case of deoxynivalenol und ochratoxin A, with the reservation that (provisional) defined reference values were assumed. The reservation referred to the fact that the data base for a sound evaluation of the effects on human health is still too limited.

The food intake must be known in order to calculate the utilization of the reference values. For this purpose, a large national study on consumption was carried out in the FRG between 1985 and 1988. With regard to the preferred foods, the amount consumed and the average body weight of the test persons were evaluated. For a differentiated presentation of the results, the test persons were divided into a total of 10 different age, sex and consumption groups, e.g., children, men, women. Among the men and women, a distinction was also made between the meat, fish and fruit eaters. The utilization of the reference value is relatively high at 34.1–82.5%. The upper value was calculated for 4–6 year old children and the lower value for women (fish eaters). Cereal products are mainly contaminated with deoxynivalenol. Ochratoxin A is also most frequently taken in by children. In addition to cereal products, especially fruit juices play a role as a source of this substance.

In comparison with the usual HPLC/UV method, it has been shown in the analysis of mycotoxins using patulin that the detection limit decreases by a factor of 100 if an isotopic dilution assay is carried out with $[^{13}C_2]$ patulin as the internal standard. After silylation and gas chromatographic/mass spectrometric measurement of analyte and isotopomer, 5.7– 26.0 μg/l of patulin were found in apple juices.

Plant-Protective Agents (PPA)

The term PPA includes all the compounds used in agricultural food production to protect cultivated plants from plant- and insect-caused diseases, parasites or weeds, or from detrimental micro-organisms. The most important groups of PPA are: (1) *herbicides* to protect the plant from weeds; (2) *fungicides* to suppress the growth of undesired fungi or molds; and (3) *insecticides* to protect the plants from damage caused by insects. In addition to these main groups, there are *acaricides* to control mites, *nematocides* to control worms or nematodes, *molluscicides* to protect the plant from snails and slugs, *rodenticides* to control rodents (mice or rats), and plant *growth regulators*. In Germany in 2003, the herbicides had the largest market share at 43%, followed by the fungicides (28%), insecticides (18%) and the growth regulators (9%). The remaining agents accounted for 2%.

The use of PPA is rewarding since it reduces losses in crop yield and stocks. It has also contributed to the control or eradication of insect- spread diseases such as malaria. Without pest control, the harvest losses of rice, which is especially susceptible, would be 24%. The use of PPA reduces rice losses to 14% and wheat, soybean and corn losses to 7–10%. Apart from losses during cultivation, about 15% of the world harvest is lost during storage due to pests in barns and silos. The substances applied must be effective against pests but safe for the user, consumer and the environment. Accordingly, these substances are evaluated with regard to their toxicological and ecological properties in the registration procedure. Their influence on, e.g., beneficial organisms, aquatic organisms, birds and mammals is tested and their degradability in the soil or in the plant is determined. Since the registration of pesticides has to be renewed at certain times and the costs of the re-evaluation have to be carried mainly by the manufacturer, only those substances which are successful on the market are put through this procedure.

PPA are applied in various forms and by various means: dusting as powder, fumigation, spraying as a liquid, or pad or furrow irrigation. The strict observance of directions for use, waiting the recommended time between final application and harvest, and restricting the application to the necessary dose, is required to maintain the residual pesticide levels in food at a minimum. Legal regulations with bans on use, stipulated maximum permissible quantities (MRL) etc. emphasize these requirements.

Contamination of food of plant origin can occur directly by treating the crop before storage and distribution (fruit and vegetable treatment with fungicides, cereal treatment with insecticides). It can occur indirectly by uptake from the soil of residual PPA by the subsequent crop, from the atmosphere or drifting from neighboring fields, or from a storage space pretreated with PPA.

Contamination of food of animal origin occurs by ingestion of feed containing stall- and barn-cleansing agents (fungicides, insecticides), by coming in contact with wooden studs and boards preserved with fungistatic agents, and veterinary medicines and, occasionally, by use of disinfected corn as fodder.

Active Agents

Insecticides

Organophosphate compounds, carbamates and pyrethroids have been used as insecticides for many years. The pyrethroids are synthetic modifications of pyrethrin I, the main active agent of pyrethrum. Pyrethrum is isolated from the capitulum of different varieties of chrysanthemums and used as a natural insecticide.

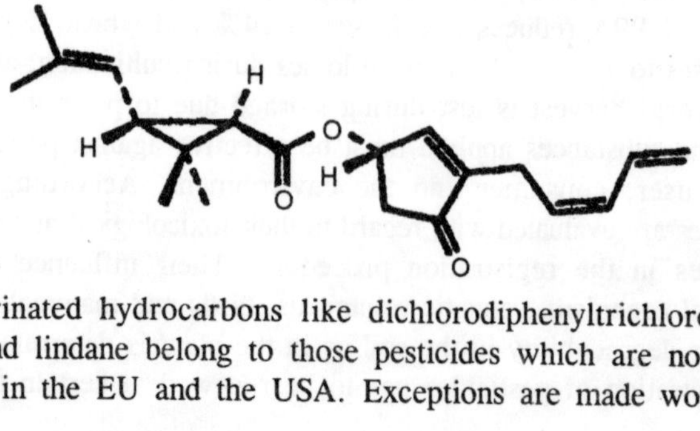

Chlorinated hydrocarbons like dichlorodiphenyltrichloroethane (DDT) und lindane belong to those pesticides which are no longer approved in the EU and the USA. Exceptions are made worldwide only for the use against mosquitoes to control malaria and only if no alternatives are available. The reason for the rejection of chlorinated hydrocarbons is their persistence. They are stable and accumulate in human beings and animals (in the fat phase) and in the environment. The half life DT_{50} (time for 50% dissipation of the initial concentration) of DDT is 4–30 years. In comparison, the DT_{50} values of the organophosphates, carbamates and pyrethroids are in the range of days to a few months, e.g., chlorpyriphos 10–120 days, carbofuran 30–60 days and deltamethrin <23 days. In Germany, the ban on DDT has resulted in a decrease in the concentration of DDT and its degradation product DDE in human milk (mg/kg milk fat) from 1.83 to 0.132.

Pest populations which are resistant to the active agents develop on longer application. Therefore, new active agents have to be continually synthesized to calculate this resistance. Examples are indoxacarb and tebufenozide.

Insecticides are mainly nerve poisons. In particular, the older active agents, e.g., parathion, chlorpyriphos and methidation are also very toxic to mammals. The toxicity falls when the ethyl groups in IXa are replaced by methyl groups. Nerve poisons inhibit acetylcholine esterase, bind to receptors which are controlled by the neurotransmitter acetylcholine or interfere with the neurotransmission in the nervous system by modifying the ion canals. Other insecticides damage the respiratory chain. As mitochondrial uncouplers, they prevent the formation of a proton gradient. Another mechanism of action is directed at the development of the pests, which can be prevented, e.g., by inhibiting the biosynthesis of chitin.

Fungicides

Real fungi (Ascomycetes, Basidomycetes, Deuteromycetes) and lower fungi (Oomycetes) can destroy entire harvests. These fungi and their spores are controlled by fungicides so that mildew, rust, leaf blight, stem rot, botrytis and other plant diseases do not occur. Depending on the mode of action, a distinction is made between the contact fungicides, which act on the surface of the plant preventing germination and/or penetration of the fungus into the plant, and systemic fungicides, which penetrate into the plant and eliminate hidden seats of disease.

The active agents can be divided into inorganic, organometallic and organic compounds. Inorganic fungicides include Bordeaux mixture, copper chloride oxide, lime sulfur and colloidal sulfur. Examples of organometallic compounds are dithiocarbamates of zinc and manganes, which are relatively often encountered as residues in foods. Most of the fungicides are, however, metal-free organic compounds.

In the case of the fungicides, too, the development of resistance necessitates the continual development of new active agents. A special innovation was the introduction of synthetic modifications of the fungal constituent strobilurin A, which has antibiotic and fungicidal activity. Examples are azoxystrobin and epoxyconazole.

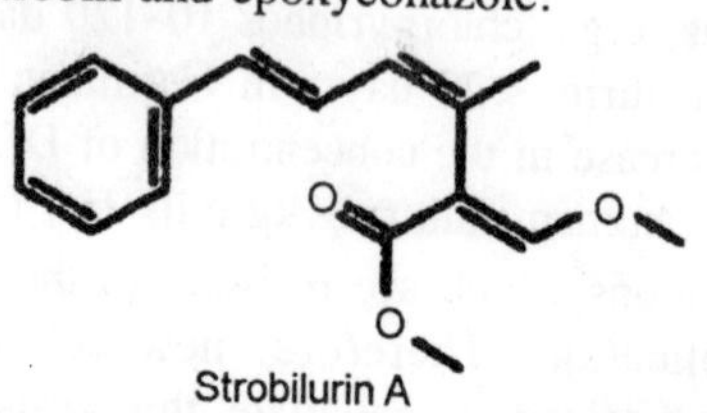

Strobilurin A

Another new class of substances are the valinamides, which has given rise to the active agent iprovalicarb. The development of active agents with increased fungicidal activity but constant relatively low toxicity for mammals has led to a considerable reduction of the dose required.

PPA residues in foods have most often been found in fruit and vegetables. Among the identified active agents, the fungicides play the biggest part.

Herbicides

A distinction is made between non-specific and specific herbicides. The former inhibit the growth of both cultivated plants and weeds. For this reason, they can only be used before sowing. The introduction of resistance genes in soybean, corn and rape seed, among others, allows their weeds to be controlled by non-specific herbicides even during growth. Selective herbicides inhibit the growth of weeds while protecting the cultivated plants. This selective action is achieved, e. g., because, unlike the weed, the cultivated plant quickly degrades the herbicide. One of the first selective herbicides was 2,4-dichlorophenoxyacetic acid, which eliminates only dicotyledon weeds but not monocotyledon cereal plants.

Newer selectively acting herbicides are the compounds amidosulfuron, mesosulfuron methyl and nicosulfuron, which belong to the class of sulfonyl ureas. Since they are very active, the amount used is very small. As in the case of the other PPA, the biochemical mechanism of action of most of the herbicides is known. They frequently target a reaction site in the chloroplasts.

Analysis

The purpose of the analyses is to detect PPA which are not registered and to expose cases where the stipulated maximum permissible amounts have been exceeded. In addition, it is necessary to continually measure the contamination of food with PPA.

The analysis of PPA residues is difficult because the number of active agents which can be taken into consideration is very large. For example, 255 compounds were registered in Germany in 2003 and maximum permissible amounts were stipulated by law for 600 compounds. However, we have to reckon with the use of about 1000 active agents worldwide. The analysis is made more difficult by the large differences in the chemical structures and the requirement for an exact quantification, the maximum permissible quantities being in the trace range, i.e., between 0.001 and 10 mg/kg. The following example gives an insight into the most important steps of a method (multimethod) with which a series of pesticides are identified.

A sample of a fruit or vegetable (ca. 10 g) is homogenized with the solvent (acetonitrile) which contains the internal standard (triphenyl phosphate). Anhydrous $MgSO_4$ and NaCl are added to bind water. After centrifugation, an SPE (solid phase extraction) sorbent is stirred in an aliquot of the supernatant to bind organic acids, pigments and sugar. After a second centrifugation, the PPA and the internal standard are identified and quantified by gas chromatography–mass spectrometry (MS). Alternatively, LC (liquid chromatography)–MS processes are also used to detect the analytes. The LC–MS process is the method of choice in the case of thermolabile active agents. Some pesticides can be accurately identified only by MS–MS measurements.

A series of PPA cannot be identified by a multi- method because their polarity and structural differences are too large. For the analysis of such active agents, special processes have been elaborated, with isotope dilution analyses also playing a part.

If metabolites of PPA are toxicologically relevant, they are also identified in the analysis, e.g., the degradation product 2,4-dimethyl-aniline together with the active agent amitraz, and dichlorodiphenyl-dichloroethane (DDD) and dichlorodiphenyldichloroethylene (DDE) with DDT.

On the plant, PPA are exposed to light and can undergo photolysis. An example is parathion. Nitrosoparathion is formed, among other compounds, and reacts with components of the cuticle. The resulting product is insoluble and is not detected by the conventional PPA analysis. It can be identified directly on the cuticle with the help of the ELISA technique.

PPA Residues, Risk Assessment

Exceeding the maximum permissible quantity

In Germany, 12,874 samples were tested for the presence of PPA in 2003. 2515 samples came from the monitoring program and 10,359 from official food monitoring. No PPA were found in 42.9% of these samples. In 50.1% of the samples, residues were present which were equal to or less than the maximum permissible quantity. Altogether 890 samples (6.9%) had residue content higher than the maximum permissible quantity.

Thus, the permissible upper limit was exceeded in more than 10% of the samples of pineapples, apricots, pears, blackberries, Chinese cabbage, dried figs, fennel, kale, currants, papaya, bell pepper, parsley, peaches, rocket and grapes. The figures for 2004 are also given for a few foods which were more contaminated.

Hydrogen cyanide, followed by bromide and chloromequat were detected in every sample of cereal. The two last mentioned active agents were found most frequently in fruit and vegetables. The maneb group, amitraz, chlorpyriphos, carbenedazine, cyprodinil and procymidones also accounted for proportions of 10% and more of the samples of these foods. The frequency of bromide is due to the fact that it is a naturally and ubiquitously occurring substance. Higher concentrations can indicate the use of bromine-containing fumigants, e. g., methyl bromide, in soil treatment or in storage.

The EU coordinated a pesticide monitoring program with the participation of Norway, Iceland and Liechtenstein in 2002. They investigated the occurrence of a total of 41 pesticides in pears, bananas, oranges/mandarins, peaches/nectarines, beans, potatoes, carrots and spinach.

Residue concentrations higher than the permissible upper limit were found most frequently in spinach (13% of the tested samples), followed by beans (7%), oranges/mandarins (4%) and peaches/nectarines (3%). Of the active agents, residues of the maneb group most often exceeded the permissible upper limit (1.19% of all the tested samples).

In the USA, a monitoring program for insecticides in 2002 showed that DDT (0.0001- 0.025 mg/kg) was the most frequently found active agent. It was found in 21% of the samples. Furthermore, chlorpyriphos methyl (17%, 0.0002–0.59 mg/kg), malathion (15%, 0.0007–0.071 mg/

kg), endosulfan (14%, 0.0001–0.166 mg/kg) and dieldrin (11%, 0.0001–0.010 mg/kg) occurred in more than 10% of the tested samples.

Risk assessment

The results of the food monitoring program in 1995–2002 were used for the risk assessment. In this time period, 30,682 samples from 130 different foods were analyzed for the presence of 160 PPA. No active substances or only traces were found in 45.7% of the samples. Concentrations exceeding the maximum permissible amounts were determined in 2.6% of the samples.

The PPA selected for the risk assessment were those which were quantified in at least 5% of the samples of three or more foodstuffs. The calculation of the food intake was based on the results of the national consumption study, which was carried out in the FRG between 1985 and 1988.

The overview of the results shows that the reference values of the selected PPA are mostly utilized to less than 1%. The dithiocarbamates are the only exception with a utilization of 7.7 to 18.3%. The lower value was found for the group of men (women 9.6%) and the upper value for children (4–6 years of age). The reason for this difference is probably the consumption of fruit because men eat less fruit than children. Fruit can contain residues of dithiocarbamates used for controlling fungal diseases.

The German Federal Office for Consumer Protection and Food Safety came to the following conclusion based on the monitoring results: "The utilization of the toxicological reference values (permissible or tolerable daily intake) is without exception very small, being only about 1% for the plant-protective agents and the persistent organochloro compounds tested. A substantial exposure was observed only with the dithiocarbamates. However, it should be taken into account that the results can be partly superimposed by the residues of naturally occurring sulfur-containing substances. The results obtained up to now do not represent the possible total contamination of the consumers with residues of plant protection agents, however a distinct section because the analysis spectrum did not include all the conceivable active substances of plant protection agents and their metabolites, but about 160 substances relevant to foodstuffs."

Natural pesticides

The natural pesticides should not be disregarded in the risk assessment. Ames and Gold (2000) have shown that in comparison with the synthetic pesticides, the pesticides occurring naturally in food are of much greater importance. Substances of this type are produced by the plant for protection against microorganisms and insects, e.g., allylisothiocyanate, benzaldehyde, caffeic acid, capsaicin, catechin, estragol, d-limonene and 4-methylcatechin.

These substances include as high a percentage of compounds which are carcinogenic as among the synthetic pesticides. It has been estimated for the USA (the conditions in Europe should be similar) that the population consumes with food on average 1500 mg per person and day of natural pesticides, but only 0.09 mg of synthetic pesticides. In summary, it can be concluded that a risk of impairment of consumer health is not discernible in the case of the proper and controlled application of registered PPA.

Veterinary Medicines and Feed Additives

The current practice in animal husbandry is the wide use of veterinary medicines, which serve not only for therapy, but to a large extent for prophylaxis and economic aims (e.g. to shorten animal growth or feeding time; to abate the risk of losses).

Veterinary preparation residues in food are ingested by humans in low amounts but continuously and, hence, could be a health hazard. This possibility was, for a long time, not carefully examined. Therefore, as in the field of pesticides, supporting and maintaining appropriate measures (imposing legally binding regulations, analytical control or supervision, elucidation of toxicological problems) has the ultimate aim of protecting human health.

Antibiotics

Antibiotics are used prophylactically to stem infections, e. g., in intensive mass animal farming and in the therapy of infectious diseases. Since they inhibit the growth of the microflora, which is present in the digestive tract of livestock, feed utilization is improved. The animals gain weight faster. This application of antibiotics as growth promoters is regarded critically in the EU and has led to bans. A constant intake of antibiotics, even at low doses, is a risk to human health since some microorganisms may become resistant and allergic reactions may develop.

Anthelmintics

In meadows and sheds, animals come into contact with their excrements and subsequently with worms in all developmental stages. Anthelmintics are used against the resulting diseases caused by worms.

Coccidiostats

The compounds of this class are added to animal feed to combat coccidiosis diseases (such as enteritis or cachexie) caused by protozoans living as parasites in intestines. Poultry and rabbits are the animals most often affected. Residues may be found in eggs.

Analysis

The aim of these analyses is:

1. to detect medicines which are banned or not approved, e.g., chloramphenicol, nitrofurans (derivatives of 2-nitrofuran, e. g., nitrofurantoin), fattening aids with estrogenic activity such as 17-estradiol, diethylstilbestrol, zeranol.
2. to check if the residue of an approved therapeutic agent is still within the permissible upper limit.

The analysis starts with screening. Antibiotics can be detected with the help of the bacteria whose growth they inhibit. Their differentiation can be further improved by means of an electrophoretic preliminary separation.

In principle, the same isolation and separation methods and mass spectrometric techniques are used for the unambiguous identification of veterinary medicines as for pesticides. Enzyme immunoassays are also used.

POLYCHLORINATED BIPHENYLS (PCBs)

The PCBs are complex mixtures of substances which have been on the market since 1950. They are widely used, e. g., as transformer oil, hydraulic fluid, heat exchange medium, dielectric fluid in condensers, plasticizer and additive for printing ink. Structural formula shows 2,2′,5,5′-tetrachlorobiphenyl and 2,2′,4,5,5′- pentachlorobiphenyl as examples.

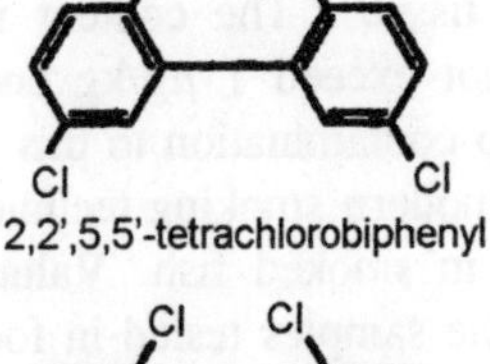

2,2',5,5'-tetrachlorobiphenyl

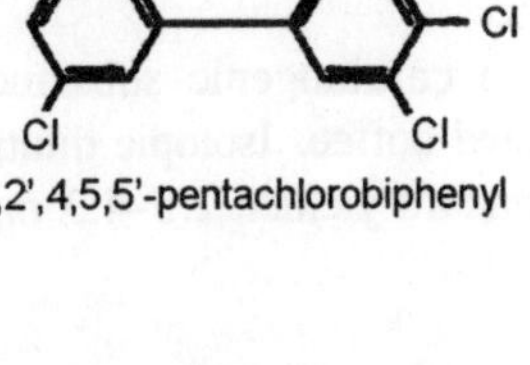

2,2',4,5,5'-pentachlorobiphenyl

As a result of their widespread use, the PCBs also came into contact with food. Because of their persistence and solubility in fat, they accumulated like in the case of DDT. Therefore, they have been increasingly identified in fatty foods since their discovery. This and the fact that PCBs can produce highly toxic dioxins in the combustion process led to the banning of the production and application of PCBs in 1989. In Germany, the contamination with PCB, e.g., in milk fat (mg/kg) has subsequently fallen on average: 0.012, 0.007, 0.003.

Harmful Substances from Thermal Processes

Polycycic Aromatic Hydrocarbons (PAHs)

Burning of organic materials, such as wood (wood smoke and its semi-dry distillation product, the wood smoke vapor phase), coal or fuel oil, results in pyrolytic reactions which yield a great number of polycyclic aromatic hydrocarbons (abouot 250 have been identified) with more than three linearly or angularly fused benzene rings, that are carcinogenic to varying extents. The quantity and diversity of compounds generated is affected by the conditions of the burning process. Benzo[a]pyrene (Bap)usually serves as an indicator compound.

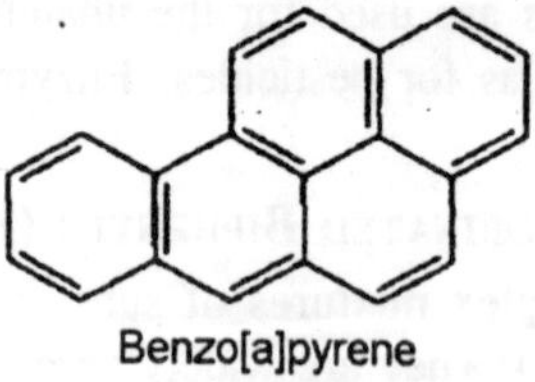

Benzo[a]pyrene

Contamination of food with polycyclic compounds can be caused by fall-out from the atmosphere (as often occurs with fruit and leafy vegetables in industrial districts), by direct drying of cereals with combustion gases, by smoking or roasting of food (barbecuing or charcoal broiling; smoking of sausage, ham or fish; roasting of coffee). PAHs accumulate in high-fat tissues. The content in meat and processed meat products should not exceed 1 μg/kg end-product measured as Bap. A reduction of Bap contamination to this limiting value has been achieved by the use of modern smoking techniques. A maximum of 5 μg/kg Bap is tolerated in smoked fish. Values less than 1.6 μg/kg were found in 95% of the samples tested in food monitoring in 2005.

Furan

Furan is possibly a carcinogenic substance. It occurs in heated food, especially in roasted coffee. Isotopic dilution analyses with $[^{2}H4]$-furan as the internal standard yielded 2.5–4.3 mg/kg furan in differently

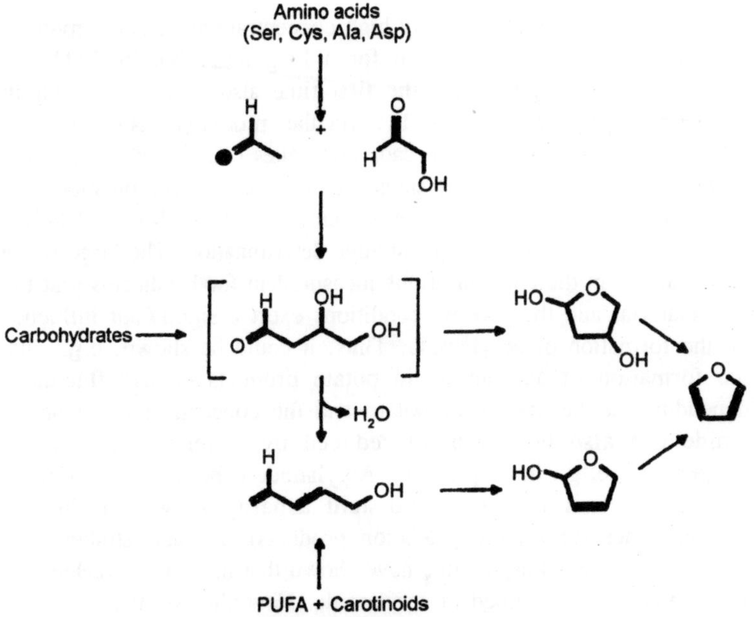

Fig. 9.2. Formation of furan on heating food.

produced coffee powders. Baby food, e.g., carrot mash and potato/ spinach mash contained 74 and 75 μg/kg respectively. Furan is formed from amino acids which yield acetaldehyde and glycolaldehyde on thermal degradation. Aldol condensation, cyclization and elimination of water are the reaction steps. Other precursors of furan are carbohydrates, polyunsaturated fatty acids and carotinoids. Furan can also be formed from the thermolysis of ascorbic acid.

Acrylamide

Polyacrylamide, produced from monomeric acrylamide (2-propenamide), has been used for decades in various industrial processes, e.g., as a flocculant in the treatment of drinking water. Especially for reasons of occupational health and safety, numerous toxicological studies on acrylamide have already been conducted. These studies have shown above all that on high exposure, acrylamide (i) binds to hemoglobin in the blood, (ii) is metabolized to reactive epoxide glycidamide and (iii) is carcinogenic on chronic exposure in animal tests. For this reason, acrylamide was put about 20 years ago into Category III A2 of carcinogenic working substances. According to EU guidelines for drinking water, the concentration of acrylamide in water should not exceed 0.1 μg/l.

The occurrence of relatively high concentrations of acrylamide in tobacco smoke has been known for a long time, but in 2002 this compound was described for the first time also as a constituent in various thermally treated foods. In particular, processed potato products such as chips, but also fine breads and cakes contain relatively high concentrations. Today, mainly stable isotope assays in combination with GC–MS or LC–MS with the use of deuterium- or carbon-13-labelled acrylamide are used for the quantitative determination. The large range of variations in the concentrations measured in food indicates that the raw material and the process conditions exert a significant influence on the formation of acrylamide. Thus, it could be shown, e.g., that the formation of acrylamide in potato products clearly fluctuates depending on the variety of potato and the concentrations of acrylamide can also be distinctly reduced by lowering the heating temperature, e. g., in deep frying. Acrylamide is preferentially formed by the reaction of the amino acid asparagine with reductive carbohydrates (or their degradation products). In fact, studies with isotopically labelled asparagine have shown that the carbon skeleton of the amino acid is retained in acrylamide. Nevertheless, the formation of acrylamide, e.g., in the case of potatoes, correlates after heating considerably better with the concentration of fructose and glucose in fresh potatoes than with the concentration of free asparagine although potatoes have a very high concentration of free asparagine at 2–4 g/kg dry weight. In the case of gingerbread, in addition to the concentration of free asparagine, the NH_4HCO_3 used as baking powder was identified as a promoter in the formation of acrylamide.

Apart from the enzymatic hydrolysis of asparagine with asparaginase, the ways in which the content of acrylamide in food can be reduced include the use of various additives, lowering the pH value and reducing the heating temperature. According to more recent calculations, the daily intake of acrylamide from foodstuffs in Germany is assumed to be about 0.57 μg/kg body weight.

Nitrate, Nitrite, Nitrosamines

Nitrate, Nitrite

The plants which belong to Group A can store very much more nitrate than those in Groups B and C, their nitrate content depending among other things on the N supplied on fertilization. Apart from the properties of the soil, even light plays a role because some plants store more nitrate when there is a lack of light. The foods of animal origin and drinking water are a further source of nitrate. It was calculated on the basis of a national consumption study that the intake of nitrate is highest in 4–6 year old children, followed by women and men who prefer fruit and vegetables in their diet rather than meat and fish. The ADI value for nitrate is utilized to 23–40% by the population.

Nitrite mainly comes from cured meat and meat products. The daily supply amounts to about 0.25 mg $NO_2^{\ominus}$.

It is remarkable that the amount of nitrate formed every day in the human organism, about 1 mg/kg body weight, is just as much as the intake in the diet. The precursor is arginine which is cleaved to give NO and citrullin. NO is oxidized to N_2O_3, which reacts with water to give nitrite. Hemoglobin oxidizes nitrite to nitrate, giving rise to methemoglobin which cannot transport oxygen. Therefore, nitrite is toxic, especially for infants (cyanosis) because their methemoglobin reductase still has low activity. This enzyme reduces methemoglobin to hemoglobin.

The toxicity of nitrate starts from the bacterial reduction to nitrite. In the human organism, about 25% of the nitrate absorbed from the food is eliminated with the saliva and up to 7% is reduced to nitrite in the mouth cavity within 24 hours by the action of bacterial nitrate reductases and transported to the stomach. About 90% of the total amount of nitrite which reaches the digestive tract comes from nitrate reduction.

The bacterial formation of nitrite led to the assumption that toxic nitrosamines can arise endogenously by the nitrosation of amines. This danger has been overestimated. Endogenous nitrosation was described as "*practically insignificant*" in the nutrition report as early as 1996.

Nitrosamines, Nitrosamides

Nitrosamines and nitrosamides are powerful carcinogens. They are obtained from secondary amines, N-substituted amides and nitrous acid:

$$RR'NH + HONO \longrightarrow RR'N{-}NO$$

$$R{-}CO{-}NHR' + HONO \longrightarrow R{-}CO{-}N(NO){-}R'$$

The nitrosonium ion, NO^+, or a nitrosyl halogenide, XNO, is the reactive intermediate:

$$NO_2^{\ominus} \underset{}{\overset{H^{\oplus}}{\rightleftharpoons}} HONO \overset{H^{\oplus}}{\rightleftharpoons} H_2ONO^{\oplus}$$

$$H_2ONO^{\oplus} \rightleftharpoons H_2O + NO^{\oplus}$$

$$H_2ONO^{\oplus} \overset{X^{\ominus}}{\rightleftharpoons} X{-}N{=}O$$

$$H_2ONO^{\oplus} \overset{NO_2^{\ominus}}{\rightleftharpoons} H_2O + N_2O_3$$

Nitrosamine formation is also possible from primary amines:

$$R{-}CH_2{-}NH_2 \xrightarrow[\text{1) Nitrosation, 2) Diazotization, 3) Deamination}]{HNO_2} R{-}CH_2^{\oplus}$$

$$R{-}CH_2{-}NH_2 \xrightarrow[\text{4) Dimerization}]{} (R{-}CH_2)_2NH$$

$$\xrightarrow[\text{5) Nitrosation}]{HNO_2} (R{-}CH_2)_2N{-}NO$$

from diamines:

$$H_2N{-}(CH_2)_n{-}CH(R){-}NH_2 \xrightarrow{HNO_2} (-N_2)$$

$$H_2N{-}(CH_2)_n{-}CH^{\oplus}(R) \xrightarrow{\text{Cyclization}}$$

$$\text{cyclo}[(CH_2)_n{-}CHR{-}NH] \xrightarrow{HNO_2} \text{cyclo}[(CH_2)_n{-}CHR{-}N(NO)]$$

R = H: diamine

R = COOH: diamino carboxylic acid

and from tertiary amines:

$$R_2N{-}CH(R^1)(R^2) \xrightarrow{HNO_2} R_2N^{\oplus}(=CHR^1R^2){-}NO$$

$$\xrightarrow{-\text{HNO}} R_2\overset{\oplus}{N}{=}CR^1R^2 \xrightarrow{H_2O}$$

$$R_2NH_2^{\oplus} + OC(R^1)(R^2) \xrightarrow{HNO_2} R_2N{-}NO$$

$$2\,HNO \longrightarrow H_2N_2O_2 \longrightarrow H_2O + N_2O$$

Nitrosamines are detected in variable amounts in many foods. The most common compound is dimethylnitrosamine, which is also a most powerful carcinogen. Some activity has been ascribed to nitrosopiperidine and nitrosopyrrolidine. In meat products cured and treated with pickling salt, 30% of the samples contained nitrosodimethylamine (NDMA; 0.5–15 μg/kg) and 13% nitrosopyrrolidine (NPYR > 0.5 μg/kg). About 25% of the cheese samples analyzed were contaminated (0.5–4.9 μg/kg).

Nitrosopyrrolidine is formed from the amino acid proline by nitrosation followed by decarboxylation at elevated temperatures, such as in roasting or frying:

$$\text{proline (pyrrolidine ring, N–H, COOH)} \xrightarrow{HNO_2} \text{N-nitrosoproline (N–NO, COOH)} \xrightarrow{-CO_2} \text{N-nitrosopyrrolidine (N–NO)}$$

The nitrosopyrrolidine (1.5 μg/kg) in meat products increases almost ten fold (to 15.4 μg/kg) during roasting and frying. An estimate of the average daily intake of nitrosamines ranges from 0.1 μg nitrosodimethylamine and 0.1 μg nitrosopyrrolidine to a total of 1 μg.

Inhibition of a nitrosation reaction is possible, e.g., with ascorbic acid, which is oxidized by nitrite to its dehydro form, while nitrite is reduced to NO. Similarly, tocopherols and some other food constituents inhibit substitution reactions. Representative suitable measures to decrease exo- and endogenic nitrosamine hazards are:

1. Decreasing nitrite and nitrate incorporation into processed meat. However, to completely relinquish the use of nitrite is a great

health hazard due to the danger from bacterial intoxication (especially by botulism).

2. Addition of inhibitors (ascorbic acid, tocopherols).
3. Decreasing the nitrate content of vegetables.

Cleansing Agents and Disinfectants

Residues introduced by large-scale animal husbandry and use of milking machines are gaining importance. The residues in meat and processed meat products originate from the surfaces of processing equipment, while residues in milk arise from measures involved in disinfection of the udder. Iodine-containing disinfectants, including udder dipping or soaking agents, may be an additional source of contamination of milk with iodine.

Also in the case of fruit and vegetables, measures have to be taken to kill pathogenic microbes. Although chlorine is normally used, it does not kill all the bacteria in the permitted concentrations. In addition, chlorine can convert pesticide residues to substances of unknown biological activity. *Ozone* is recommended as an alternative. In disinfection, it is 1.5 times as active as chlorine, kills micro-organisms which are not attacked by chlorine and destroys pesticide residues. It decomposes to molecular oxygen with a half-life (aqueous ozone solution) of 20 minutes at room temperature. To disinfect process water, 0.5–5 mg/kg of ozone are used for <5 minutes. Apart from disinfection, ozone also delays the ripening of fruit.

Polychlorinated Dibenzodioxins (PCDD) and Dibenzofurans (PCDF)

Polychlorinated dibenzo-p-dioxins (PCDD) and polychlorinated dibenzofurans (PCDF), informally called "dioxins", occur as companion compounds or impurities in a large number of bromine- and chlorine-containing chemicals.

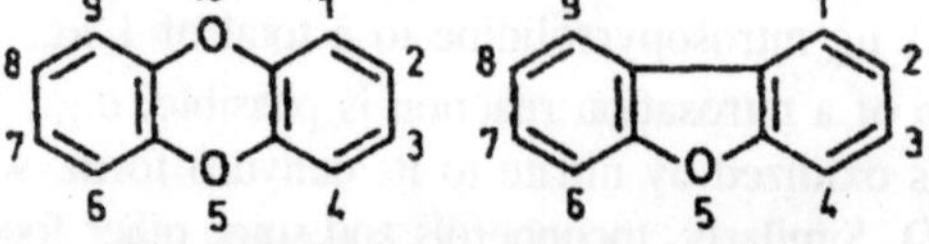

Furthermore, they are formed in many thermal processes (600 °C > T ³ 200 °C) in the presence of chlorine or other halogens in inorganic or organic form. Consequently, they are widely distributed in the environment. The number of isomers (congeners) is large, For rodents, 2,3,7,8-tetrachlorodibenzodioxin (2,3,7,8-TCDD, "Seveso dioxin") has proved to be expecially toxic (LD_{50} = 0.6 μg/kg, guinea-pig) and carcinogenic. With the exception of PnCDD, the toxicity of other congeners is lower and is generally expressed as toxicity equivalent factors (TEFs), based on 2,3,7,8-TCDD (TEF = 1). With the help of these values, 2,3,7,8-TCDD equivalents (TCDD equivalents, TEQ) can be calculated, which are a measure of the total exposure to corresponding compounds.

The daily intake of dioxin in industrial countries is estimated at 1–3 pg TEQ/kg body weight. The half life and the absorption rate are assumed to be 7.5 years and 50% respectively. In 1997, the WHO set a value of 1–4 pg TEQ per kg body weight and day as a tolerable, lifelong intake (tolerable daily intake, TDI). Due to the concentration of dioxins in the fat phase, mother's milk in the industrial countries has on average 10–35 pg TEQ per g of fat and in the developing countries, 10 pg TEQ per g of fat.

INDEX